Lecture Notes in Computer Science 5373

Commenced Publication in 1973
Founding and Former Series Editors:
Gerhard Goos, Juris Hartmanis, and Jan van Leeuwen

Michela Bertolotto   Cyril Ray   Xiang Li (Eds.)

Web and Wireless Geographical Information Systems

8th International Symposium, W2GIS 2008
Shanghai, China, December 11-12, 2008
Proceedings

 Springer

Volume Editors

Michela Bertolotto
School of Computer Science and Informatics
University College Dublin, Ireland
E-mail: michela.bertolotto@ucd.ie

Cyril Ray
Naval Academy Research Institute
Brest, France
E-mail: cyril.ray@ecole-navale.fr

Xiang Li
Key Lab of Geographic Information Science
Ministry of Education
East China Normal University
Shanghai, China
E-mail: xli@geo.ecnu.edu.cn

Library of Congress Control Number: 2008940425

CR Subject Classification (1998): H.2.8, H.3.5, H.2.8, H.5

LNCS Sublibrary: SL 3 – Information Systems and Application, incl. Internet/Web
and HCI

ISSN   0302-9743
ISBN-10   3-540-89902-2 Springer Berlin Heidelberg New York
ISBN-13   978-3-540-89902-0 Springer Berlin Heidelberg New York

Springer is a part of Springer Science+Business Media

springer.com

© Springer-Verlag Berlin Heidelberg 2008
Printed in Germany

Typesetting: Camera-ready by author, data conversion by Scientific Publishing Services, Chennai, India
Printed on acid-free paper   SPIN: 12578207   06/3180   5 4 3 2 1 0

Preface

The 8th edition of the International Symposium on Web and Wireless Geographical Information Systems (W2GIS 2008) was held in December 2008, in the vibrant city of Shanghai, China. This annual symposium aims at providing a forum for discussing advances on recent developments and research results in the field of Web and wireless geographical information systems. Promoted from workshop to symposium in 2005, W2GIS now represents a prestigious event within this dynamic research community. These proceedings contain the papers selected for presentation at this international event.

For the 2008 edition, we received 38 submissions from 16 countries. All submitted papers were related to topics of interest to the symposium. Each paper received three reviews. Based on these reviews, 14 papers were selected for presentation and inclusion in the proceedings. The accepted papers are all of excellent quality and cover topics that range from mobile networks and location-based services, to contextual representation and mapping, to geospatial Web techniques, to object tracking in Web and mobile environments.

We wish to thank all authors that contributed to this symposium for the high quality of their papers and presentations. Our sincere thanks go to Springer's LNCS team. We would also like to acknowledge and thank the Program Committee members for the quality and timeliness of their reviews. Finally, many thanks to the Steering Committee members for providing continuous support and advice.

October 2008

Michela Bertolotto
Cyril Ray
Xiang Li

Organization

Symposium Chairs

M. Bertolotto	University College Dublin, Ireland
C. Ray	Naval Academy Research Institute, France
X. Li	East China Normal University, China

Steering Committee

M. Bertolotto	University College Dublin, Ireland
J.D. Carswell	Dublin Institute of Technology, Ireland
C. Claramunt	Naval Academy Research Institute, France
M. Egenhofer	NCGIA, USA
K.J. Li	Pusan National University, Korea
T. Tezuka	Ritsumeikan University, Japan
C. Vangenot	EPFL, Switzerland

Program Committee

A. Abdelmoty	Cardiff University, UK
S. Anand	University of Nottingham, UK
M. Arikawa	University of Tokyo, Japan
T. Badard	Laval University, Canada
A. Bouju	University of La Rochelle, France
T. Brinkhoff	IAPG, Germany
K. Clarke	University of California, Santa Barbara, USA
M.L. Damiani	DICO - University of Milan, Italy
R.A. de By	ITC, The Netherlands
M. Duckham	University of Melbourne, Australia
A. Edwardes	University of Zurich, Switzerland
R. Güting	Fernuniversität Hagen, Germany
B. Huang	Chinese University of Hong Kong
Y. Ishikawa	Nagoya University, Japan
C.S. Jensen	Aalborg University, Denmark
H.A. Karimi	University of Pittsburgh, USA
B. Kobben	ITC, The Netherlands
J. Komarkova	University of Pardubice, Czech Republic
Y.J. Kwon	Hankuk Aviation University, Korea
S. Li	Ryerson University, Canada
M.R. Luaces	University of Coruna, Spain
H. Martin	University of Grenoble, France

G. Mc Ardle	University College Dublin, Ireland
P. Muro-Medrano	Universidad de Zaragoza, Spain
S. Mustiere	IGN, France
S. Nittel	University of Maine, USA
B.C. Ooi	National University of Singapore
G. Percivall	Open Geospatial Consortium, USA
D. Pfoser	Computer Technology Institute, Greece
P. Rigaux	University of Paris-Dauphine, France
M. Schneider	University of Florida, USA
C. Shahabi	University of Southern California, USA
S. Spaccapietra	EPFL, Switzerland
K. Sumiya	University of Hyogo, Japan
Y. Theodoridis	University of Piraeus, Greece
M. Tomko	Universtiy of Zurich, Switzerland
M. Ware	University of Glamorgan, UK
R. Weibel	University of Zurich, Switzerland
S. Winter	The University of Melbourne, Australia
O. Wolfson	University of Illinois at Chicago, USA
A. Zipf	University of Bonn, Germany

Local Industrial Chair

| L. Ye | Shanghai Galileo Industries, China |

Local Organizing Committee

L.Z. Yu	East China Normal University, China
J. Shu	East China Normal University, China
L. Yu	Shanghai Ubiquitous Navigation Technologies Ltd., China
J.P. Wu	East China Normal University, China
M.H. Ji	East China Normal University, China
Y.F. Wang	East China Normal University, China

Sponsoring Institutions

East China Normal University
Shanghai Ubiquitous Navigation Technologies Ltd.
The Geographical Society of Shanghai
Foundation 985 Platform for Urbanization and Public Safety

Table of Contents

Indoor Space: A New Notion of Space

Ki-Joune Li

Department of Computer Science and Engineering
Pusan National University, Pusan 609-735, South Korea
lik@pnu.edu

1 Introduction

With the rapid progress of GIS and ubiquitous computing technologies, the space that we are dealing with is no longer limited to outdoor space but being extended to indoor space. Indoor spaces has a number of differences from outdoor space, which make it difficult to realize GIS for indoor space. In order to extend the scope of GIS to indoor space, and provide integrated and seamless services, it is required to establish new theories, data models, database management systems for indoor spatial data, and applications for indoor space. In this paper, we investigate the research issues of indoor space and indoor GIS.

2 Indoor Spatial Data Model and Spatial Theory

The major differences of the indoor space from the outdoor space include the constraints of space. No constraint is in Euclidean outdoor space, while the natures of indoor space are determined by the constraints of architectural components, such as doors, corridors, floors, walls, and stairs. In order to analyze the indoor space, the description and understanding of these architectural components are major tasks and we need a data model to describe the constraints in indoor space. In architectural engineering communities, several data models have been proposed for this purpose and one of which is IFC (Industry Foundation Classes)[1] proposed by the IAI (the International Alliance for Interoperability)[2]. However, this model is focused on the architectural engineering aspects, such like construction management and facility management rather than spatial information services. An indoor spatial data model for GIS view point is suggested to describe the interior 3D space as the LOD 4 of CityGML [3], which is an international standard of OGC. This model is much closer to the data model for GIS than IFC but mainly intended to the visualization of indoor space rather than indoor spatial services and analysis, which may be complicated like evacuation routing analysis.

An important requirement for indoor spatial data model is related with the notion of *cellular space* (or *symbolic space*[4][5]). While a region query in outdoor space is given with coordinates such as (x_1, y_1) and (x_2, y_2), the query in indoor space is often based on cellular notation like "How many person are in room 422?". The room number of this example is a cell identifier, which differs from

M. Bertolotto, C. Ray, and X. Li (Eds.): W2GIS 2008, LNCS 5373, pp. 1–3, 2008.

the coordinates in Euclidean space. For example, the location in a train, which belongs to indoor space, is not identified by its coordinates but by the wagon and seat numbers.

While the Euclidean space has geometric and topological properties, the cellular space has only topological properties, which are to be explicitly specified. Among several types of topology in indoor space, the connectivity between indoor space and outdoor space should be seriously considered as well as the topology between indoor cells. For example, the entrance of a building is an important topology to connect an indoor space with outdoor space and provides a seamless service such as seamless navigation service.

Consequently the spatial information described by cell identifier should be differently stored, managed and processed. Certain space theories and data models developed for GIS of outdoor space must be replaced with new ones [6] due to the difference between cellular space and Euclidean space.

3   Integration with Indoor Positioning Technologies

While the position in an outdoor space is relatively easily collected (e.g. by GPS), the indoor positioning technologies are unfortunately immature. No single technology provides a stable positioning method but hybrid approaches are being considered for indoor positioning. And several aspects of indoor space are closely related with the indoor positioning technology. For example, when RFID technology is used for the indoor positioning, the granularity of cell is differently defined from other technologies and tracking methods for RFID are therefore different. Other parts of GIS should be tuned for RFID technique.

4   Indoor Spatial Database Management Systems

Since the properties of indoor space are different from those in outdoor space, the management systems for indoor spatial databases have different functionalities, which are summarized as follows;

- Representation, storage, indexing, and query processing for cellular spatial databases, 3D geometric databases, and moving objects in indoor space.
- Integration of cellular space and Euclidean space,
- Realtime tracking of moving objects [8],
- Continuous query processing for moving objects, and
- Indoor spatial analysis [7].

5   Conclusion

There has been an evolution of space on GIS from macro space to micro space and the indoor space is considered as a micro space compared with the conventional spaces. In order to realize indoor GIS, we have several challenges from theoretical background of indoor space and indoor spatial data models to database management systems and applications for indoor space. The research on indoor GIS stays at the beginning step. However we are expecting a significant progress on indoor GIS in a near future, especially combined with other related technologies such as ubiquitous computing.

References

1. IFC Official Web Page, http://www.iai-tech.org/products/ifc_specification
2. IAI Official Web Page, http://www.iai-international.org
3. CityGML Official Web Page, http://www.citygml.org
4. Becker, C., Durr, F.: On Location Models for Ubiquitous Computing. Journal of Personal and Ubiquitous Computing 9, 20–31 (2005)
5. Kolbe, T.H., Groger, G., Plumer, L.: CityGML – Interoperable Access to 3D City Models. In: The proceedings of the 1st International Symposium on Geo-information for Disaster Management. Delft, Nederlands (2005)
6. Lorenz, B., Ohlbach, H.J., Stoffel, E.-P., Rosner, M.: Towards a semantic spatial model for pedestrian indoor navigation. In: Hainaut, J.-L., Rundensteiner, E.A., Kirchberg, M., Bertolotto, M., Brochhausen, M., Chen, Y.-P.P., Cherfi, S.S.-S., Doerr, M., Han, H., Hartmann, S., Parsons, J., Poels, G., Rolland, C., Trujillo, J., Yu, E., Zimányie, E. (eds.) ER Workshops 2007. LNCS, vol. 4802. Springer, Heidelberg (2007)
7. Li, D., Lee, D.L.: A Topology-based Semantic Location Model for Indoor Applications. In: The Proceedings of ACM GIS (to appear, 2008)
8. Liao, L., Fox, D., Hightower, J., Kautz, H., Schulz, D.: Voronoi Tracking: Location Estimation Using Sparse and Noisy Sensor Data. In: The proceedings of the IEEE/RSJ International Conference on Intelligent Robots and Systems (2003)
9. Raubal, M.: Agent-Based Simulation of Human Wayfinding: A Perceptual Model for Unfamiliar Buildings, Ph.D Thesis, Vienna University of Technology, Vienna (2001)

HyperSmooth: A System for Interactive Spatial Analysis Via Potential Maps

Christine Plumejeaud[1], Jean-Marc Vincent[1], Claude Grasland[2],
Sandro Bimonte[1], Hélène Mathian[2], Serge Guelton[1], Joël Boulier[2],
and Jérôme Gensel[1]

[1] Laboratoire d'Informatique de Grenoble, BP 72, 38402 Saint-Martin d'Hères, France
`{firstname.name}@imag.fr`
[2] UMR Laboratoire Géographie-Cités, 13 rue du four, 75006 Paris, France
`{firstname.name}@parisgeo.cnrs.fr`

Abstract. This paper presents a new cartographic tool for spatial analysis of social data, using the potential smoothing method [10]. The purpose of this
method is to view the spread of a phenomenon (demographic, economical, social, etc.) in a continuous way, at a macroscopic scale, from data sampled on
administrative areas. We aim to offer an interactive tool, accessible through the
Web, but ensuring the confidentiality of data. The biggest difficulty is induced
by the high complexity of the calculus, dealing with a great amount of data. A
distributed architecture is proposed: map computation is made on server-side,
using particular optimization techniques, whereas map visualization and param-
eterisation of the analysis are done with a web-based client, the two parts communicating through a Web protocol.

Keywords: multiscalar spatial analysis, potential maps, interactive maps, spatial decision support system.

1 Introduction

Recent advances in the Web domain have led to new research issues in interactive and
dynamic cartography (Cartographic Web) [8]. In this context, the interdisciplinary
research group *HyperCarte* which gathers researchers in Geography, Statistics and
Computer Science works on the design and development of a set of interactive spatial
analysis tools for the representation and the analysis of social, economic and environmental phenomena. Cartographic Web technologies represent a possible approach
as they grant flexibility and interactivity. Indeed, spatial analysis consists in the exploration of spatial data in order to formulate, compare and validate hypotheses. The
strong link between interactivity, exploration and data analysis is described by
MacEachren [13] through the "map-use cube" which represents visually the degree of
interactivity, the type of target and the degree of data knowledge necessary for the
user in the *Exploration, Analysis, Synthesize and Presentation* steps of the spatial
decision making process. A map is interactive if it gives access to other data [1], [11].
For instance, through a simple click on a part of an interactive map, a new piece of

M. Bertolotto, C. Ray, and X. Li (Eds.): W2GIS 2008, LNCS 5373, pp. 4–16, 2008.

information (another interactive map, a multimedia document, etc.) can be accessed. There is no complex query language to be mastered by the user. Thanks to this, one can touch a wider public, allowing non-computer science aware people, such as policy makers, geographers, statisticians, stake holders, and so on, to create potential maps.

This paper describes *HyperSmooth* a tool based on the Web, which generates dynamically continuous interactive maps using administrative, environmental, or economical data, collected on grids or territorial meshes. HyperSmooth implements a smoothing method, called the *potential transformation method* [10], which provides multiscale cartographic representations, abstracting the real observed data. More specifically, the goal is to visualize and analyze the spatial distribution of socio-economic phenomena at a macroscopic level. So far, no Geographic Information System (GIS) [12] and Exploratory Spatial Analysis tool [6] offer an interactive analyzis of true potential maps. The main issues of this spatial analysis method and of its cartographic representation are its high computing cost, which generally hampers interactivity. To overcome this problem, Hypersmooth distributes the computation on a multiprocessor server, which performs a parallelization of calculation tasks, making possible the visualization of maps for a web interactive client with a secured connection.

The paper is structured as follows. Section 2 presents the research motivations that have led to HyperSmooth. The potential transformation method is described in section 3. Our prototype is presented in section 4, and section 5 shows the results of our approach. Conclusion and future work are detailed in section 6.

2 Research Motivations

2.1 Using Exploratory Spatial Data Analysis Systems

Geographic Information Systems (GIS) allow storing, visualizing and analyzing spatial data [12]. Spatial data can then be analyzed by means of geostatistic and classical analysis tools. Commercial GIS implement several vectorial data analysis tools (i.e. buffer, overlay, etc.). Some interpolation methods to create continuous maps (i.e. Inverse Distance Weighted (IDW), Krigging, Spline Polynomial Trend) have also been implemented. Yet it has been recognized that GIS are not Spatial Decision Support Systems (SDSS) [7]. SDSS helps decision makers to solve spatial decisional problems by providing a simple, interactive and flexible interface, managing aggregated data, handling complex spatial data structures (i.e. spatial hierarchies, field data, networks, etc.), and granting effective response times. Different kinds of SDSS have been developed. For instance, Visual Spatial Data Mining systems [3] integrate spatial data mining and GIS functionalities, Spatial OLAP tools add GIS functionalities to OLAP systems [2], and the Exploratory Spatial Data Analysis tool (ESDA) [6] allow exploring vectorial and field data by means of interactive maps and graphic displays (scatter plots, histograms, parallel coordinates). In these systems, users can interact with the map and trigger spatial analysis operations. Spatial analysis tools for field data implemented in SDSS systems offer many functions such as: summarization (i.e. data is aggregated for each cell of the grid map), reclassification (i.e. data value is transformed using data mining algorithms and/or statistical methods, etc.), change of resolution of a grid, user-defined weighted point transformations, triangulation or detection of polygons, and calculation of weighted densities [1]. SDSS systems provide adequate user interfaces to provide the best insight into the spatial data set, as well as particular computation methods to speed-up calculation times. However, to the best of our knowledge no GIS nor ESDA implement the method of potential described in the following section.

2.2   The Method of Potential Versus Classical Interpolation Methods

For the purpose of spatial analysis, geographers need to give a continuous representation of data (currently collected on grids), in order to abstract from the initial grid built during the collect of these data. The method of potential is an innovative approach for such problems, proposed by our HyperCarte research group [10]. Many commercial GIS's offer rough smoothing methods, which produce a continuous map according to the density of each territorial unit. Various methods have been proposed: triangulation with linear interpolation, kriging, polynomial estimation, Shepard, etc. These methods generally rely on the assumption that the density z_i of each territorial unit can be located at the centroïds (x_i, y_i) of those units and they interpret them as a sample of points of a surface $z=f(x,y)$, the equation of which has to be estimated. The problem is that the sample $\{z_1, \ldots z_i, \ldots z_n\}$ defined as the density at the barycenter of territorial units mixes heterogeneous neighbourhoods because units have different sizes and shapes. Consequently, it is not surprising to observe that these interpolation methods produce distinct maps of density when they are applied to various territorial divisions having the same area but different shapes, while they describe the same information. This non-convergence of results can be explained by the lack of control on the scale of generalisation, or more precisely by an imperfect filtering of the harmonics of the distribution of population density [16]. This problem is described like the Modifiable Area Unit Problem (MAUP) or Change Of Support Problem (COSP) and some solutions have been proposed and argued in the scientific literature [9], [14].

The method of the potential is in fact related to more general statistical concepts (non parametric estimators of kernel density), and its main advantage relies precisely on the possibility to introduce a scale parameter which give the opportunity to control the spatial uncertainty of the aggregated information [10]. For that reason, the method of potential can be used in order to solve the problem of instability of the results, particularly the correlations, according to the chosen spatial resolution. In order to explore and analyze geographic phenomenon at multiple scales, there is a need for an ESDA system implementing the method of potential, providing a user-friendly interactive interface. Due to the high cost of the computation, this system should ensure high calculus performances. This system should also protect the user's data.

3   Potential Transformation Method

This section presents the principles of the potential transformation method highlighting the complexity of its implementation, but also the great opportunities that it offers for an explorative cartography of the spread of socio-economic phenomenon.

A continuous cartographic representation of discrete spatial phenomena is necessary when an abstraction of the spatial grid is needed because the data grid is heterogeneous, or because the cell level has simply no meaning with regard to the analyzed phenomenon. This allows for a spatial distribution of the phenomenon without any reference to the underlying administrative territorial subdivision.

The method handles the geographic space through a grid composed of irregular territorial units. Different kinds of indicators are associated with each cell of the grid: number of inhabitants, plants, cars, the quantity of wealth expressed by the GDP, etc. The territorial grid can be, for instance, the municipality territorial subdivision, and this sort of grid is usually nested in an upper level grid (like the departmental grid), that is to say that cells of the lower grid are grouped to form a bigger cell in the upper level; then the sum of the values of an indicator associated to the lowest cells is equal to the value of the formed cell in the superior grid (this is the additive property of the used indicators). For each indicator, the method calculates the value of the potential in each point (or cell) of the discretized space. Discretization is a subdivision of space into regular plots using, for instance, a projection into a geometric space. In every location of the geometric space, the calculated potential must be understood as the likely value for the considered indicator, depending on the contribution of each cell of the geographic space weighted by the distance to this location.

Let A be the set of territorial units, a an element of this set, Sa the indicator value associated to the unit. Knowing that indicators are additive, and that their contribution is proportional to the distance δ between a and the point M, we define the potential $\Phi(M)$ for a point M of the geometric space as:

$$\Phi(M) = \sum_{a \in A} S_a f(\delta(a, M)) . \tag{1}$$

For instance, if A is a set of European municipalities, and S the number of centenarians living there, we wish to estimate the potential value of these people in every location M of Europe. For each city a, Sa is the number of centenarian inhabitants, and g_a is the centre of the municipality (i.e. the centroid, its administrative or economic centre, etc.). The distance $\delta(a,M)$ is defined as the distance d existing between a location M and g_a. Then, the contribution of each element a of A to the potential value in M is weighted by a function f of the distance d, because the effect of an indicator usually decreases with distance: it is maximal at zero distance, and minimal at an infinite distance. To define correctly the potential in function of a parameter, normalization in each point O of the space is applied using [2]:

$$\int_{R^2} f(d(O, M)).dM = 1 \text{, is also } \int_A \Phi(M).dM = \sum_{a \in A} S_a . \tag{2}$$

The total sum of the indicators is equal to the integral of the potential. We obtain a redistribution of the mass of the considered space. Using the gravity model metaphor, $\Phi(M)$ can be interpreted as the attraction of the environment on a mobile point situated in M, whose displacement vector is $-grad$ Φ. Another possible interpretation is that $\Phi(g_a)$ measures the influence of a mass situated at point g_a on the set of points M of its neighbourhood. By example the centenarians living in Nuoro, a city in Sardinia, will contribute much more to the estimation of a point M situated in its neighbourhood than the centenarians living in Rome. From a methodological point of view, this approach is quite similar to signal treatment processing based on the de-convolution of the sampled signal. The computational cost of the equation [1] depends on the administrative grid size (the number n of elements of A) and on the resolution of the image that represents the map (the number m of estimated points M).

The calculation depends mainly on the function f, called **spatial interaction function**. The shape of the function integrates the hypothesis made about the modalities of the spatial diffusion of the studied phenomenon. Three models of functions are proposed: a model with a limited support (*Disk*), an exponential model for close interactions (*Gaussian*), and a model with a long scope (*Exponential*). The last method allows, for instance, to model and study the propagation of human epidemics: their diffusion could be represented on long or short distances, according to the mobility range of the contaminated element (human, animal or plant), depending on its conveyance. Users can test the different kinds of model by applying different functions. The analysis of the phenomenon depends also on the scope p of the interaction function. The scope is defined as the average distance action of a mass on its neighbourhood. It is linked to the form of the interaction function by the equation [3].

$$p = \int_{R^2} d(O,M) f(d(O,M)).dM = \int_0^{+\infty} f(r)2\pi r^2.dr. \qquad [3]$$

The scope could be interpreted as the spatial scale of the study chosen for the representation. The couple (function, scope) supports the economic and sociological hypothesis regarding the spatial diffusion of the studied phenomenon. For an exploratory approach, the ability to choose the values of this couple (function, scope) is crucial for the user, who would like to modify interactively these two parameters according to the resulting map. For instance, the figures 1 and 2 show two potential maps representing the population density, made with Gaussian function on scope of 250 km and 1000 km respectively. By comparison, the map of figure 1 is more detailed and shows the high-density population areas more precisely than the map of figure 2 that, on the contrary, generalises the phenomenon.

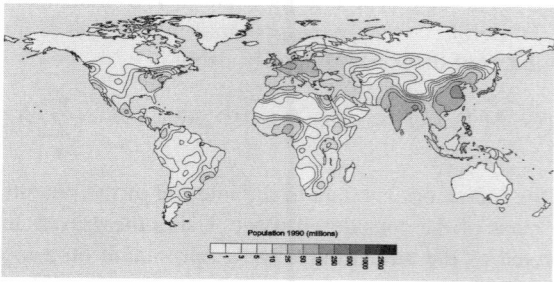

Fig. 1. Potential map of population in the World, Gaussian function, with a scope of 250 km

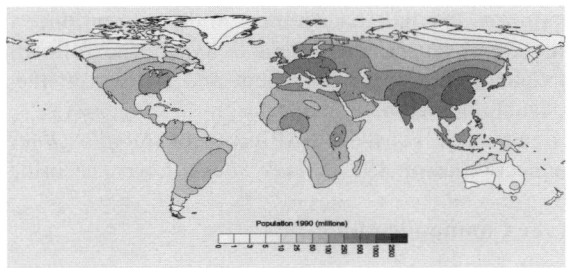

Fig. 2. Potential map of population in the World, Gaussian function, with a scope of 1000 km

In addition, the type of the distance used for computation is not trivial. For instance, it can depend on the size of the space covered by the map: at a continental scale, it is not possible to use the Euclidian distance without introducing bias. The orthodromic distance is then a better choice because it takes into account the sphericity of the planet. However, the analyzed phenomenon could necessitate other distances taking account of the anisotropy of the space, such as the transport distance expressed in terms of the time duration required to reach two remote locations.

4 Design of an Interactive Web-Based System Computing Potential Maps

In this section, we present our prototype, *HyperSmooth*, for the calculation and the visualization of potential maps. HyperSmooth permits, through particular optimization techniques, to rapidly compute potential (continuous) maps, and thus offers a high degree of interactivity, which is required for spatial analysis.

HyperSmooth is based on a client-server architecture with a Web-based client and a cluster of computation stations (see figure 3). The computation of potential maps requires a huge volume of processing resources. The Java client achieves the calculations for the cartographic visualization, while the server is responsible for the heavy computation of potential maps. At this stage, the computation only uses the orthodromic distance.

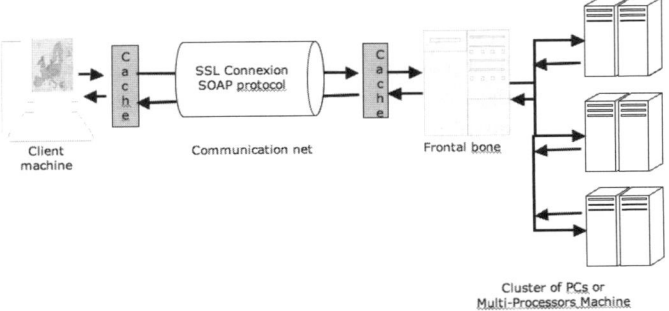

Fig. 3. The distributed architecture of HyperSmooth

The server optimizes the data management in order to produce intermediary results in a few seconds which guarantees interactivity. In the following subsections, we detail the communication between the client and the server, the optimization techniques, and the visual interactive user-interface.

The server is accessible via the SOAP protocol (*Simple Object Access Protocol*) with a secure connection using SSL (*Secure Socket Layer*) securing confidential data.

4.1 Client-Server Communications

A secured access to the application is mandatory because users visualize potential maps made with their own data, that they do not necessarily want to disseminate, and it is possible to calculate original confidential data using potential maps. Then the communication between the server and the client is realized by means of the protocol SOAP [17] for its security, portability and accessibility properties. Indeed, SOAP can use HTTP and HTTPS ports, which usually are not filtered, and it can be coupled with SSL, which provides the encryption and security. Thanks to HTTP(S), users can get a secured access on the service anywhere at anytime. Moreover, Web services grant interoperability to the server side, making HyperSmooth an extensible framework where new GIS or SDSS clients could be plugged on to the web service offered by the server side.

The potential grids calculated by the server depend on various parameters (the resolution, the framing, the interaction function and the scope), which are specified by the client to the server. The exchanges between the two parts are limited to those elements, and then the client manipulates the grid data to build the image. Indeed, the client is in charge of the customisation of the map to comply with user preferences, such as the color palette, the number of classes and the type of progression for the distribution of the colors. The client can generate reports (text and/or HTML files) on the fly. These reports contain the geographic coordinates of each point M, together with its potential value. The client saves these files (that is to say the grid data, including the calculus parameters) making possible the rapid repainting of the image when the graphic preferences are changed.

4.2 Optimization Techniques

Caching techniques are not adapted to the potential problem because each query has to generate a global result, which cannot be pre-calculated as it depends on the parameters of the analysis. However, a detailed analysis of the calculation tasks performed on the server's side shows that there are redundancies which can be avoided to optimize some parts of the calculation. Indeed, two problems arise when computing the potential $\Phi(M)$ (cf. equation [1]): the sum is applied to a huge number of elements, and the cost of the orthodromic distances $d(M, ge)$ calculus is high because it requires the computation of arccosine, cosine and sine angles.

The sum is reduced using a *cut-off* algebraic method. Contrary to the geometric cut-off which reduces the calculation to a fixed radius, our method takes into account some distant points whose weight Sa (statistic value) acts upon the result of the calculation. The cut-off algorithm exploits the organisation of data into a quad-tree. A point M with its coordinates and its indicators are associated to each leaf of this tree. The

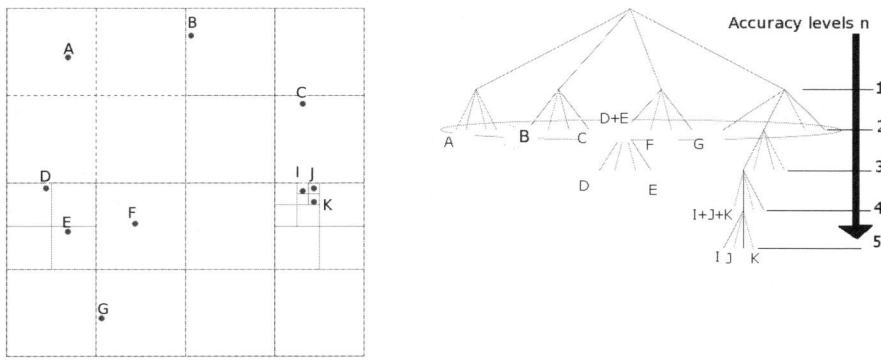

Fig. 4. An example of tree construction from a discretized study space

depth of the tree is n. Each point is then summed by groups of 4 elements (the neighbours in the grid) – see figure 4. The calculation of the potential $\Phi(M)$ is called recursively summing the product between the indicators associated to the leaves and their distance from the point M. The visit to each branch of the level $n\text{-}1$ depends on the following test, which checks if the weight of the children of the $node_{n-1}$ is negligible or not:

$$(\sum_{node_{n-1}} S_e) * d_{min} \leq \varepsilon * \Phi(M)_{summed}. \qquad [4]$$

A crucial step of the algorithm is the definition of the epsilon value in order to not discard too many points. By default this value is 1/1000 of the sum of the indicators.

The evaluation of formulas with terms in arccosine, sine, cosine slows down the calculus, but a tabulation of these functions allows the problem to be overcome. Indeed, the values of these functions are pre-computed in fine and regular subdivisions of a given interval, and then, the algorithm estimates the value of a certain angle with the nearest pre-computed angle value. The number of subdivisions of this interval can be chosen.

4.3 Interactive Visualization

The user interface is composed of a main panel containing the interactive map (Data Panel) and several windows that enable the user to parameterize the spatial analysis (Control Panel). He can select the dataset he wishes to work with in the "Control Panel", see figure 5-2. There are two combo boxes listing available datasets, one for the numerator and the other one for the denominator of the potential ratio.

The specific parameters for the computation of potential maps are: the type of the interaction function to be applied, the resolution of the computed grid and the scope. All these parameters can be selected through the panel shown on figure 5-1. There are four types of functions listed in a combo-box: *gaussian*, *exponential*, *disk* and *amortized disk*. The resolution is expressed like the number of cells in width and height of the computed grid, in a free text field. The average scope is an ordinal value expressed in kilometres, which can be entered either in a text field, or using a cursor on a slide bar.

The cartographic representation of the potential map is paint inside a tab of the Data Panel (see figure 5-5). The visualization of a potential map is obtained by clipping the raster image on the vector map of the study space. The raster image colour graduation reflects the intensity of the phenomenon, and can be customized for each map. The Data Panel contains also three panels for the colour palette, the type of colour distribution, and the number of colour classes (see figure 5-3 and figure 5-4).

Through the Control Panel functionalities, the user can tune the spatial analysis and the results will be automatically refreshed in the Data Panel. Moreover, the client redefines classical navigation map facilities: zoom and pan. These functions trigger the computation of potential maps whose results are quickly returned. Thus, the cartographic component is composed of interactive maps that are computed in response to the user's actions in effective times allowing exploration and analysis of spatial continuous phenomena.

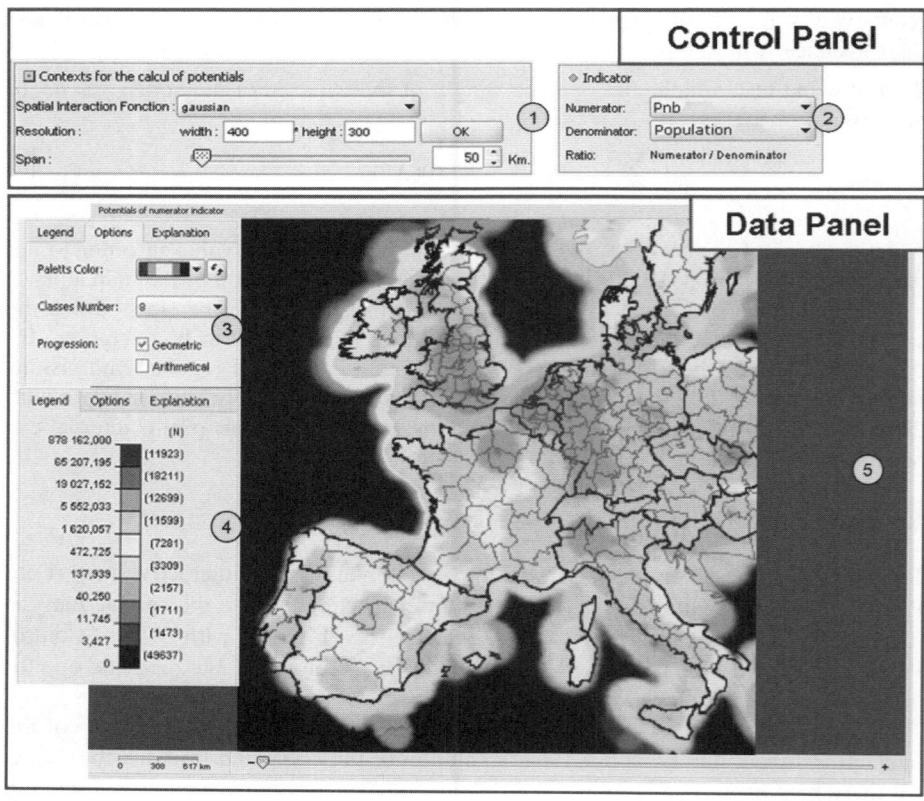

Fig. 5. The HyperSmooth user interface. Control Panel: 1) Spatial interaction function, scope and resolution, 2) Dataset to explore, Data Panel 3) Options, 4) Legend, 5) Interactive grid map.

5   Experiments: Population Distribution at Different Scales

In this section we present the experiments conducted on the dataset of the European population counted at the municipal level, which contains 116203 geographical units. These data have been analyzed using HyperSmooth to find a measure of the local polarization in Europe using a meager information (population and surface of cities on 1999) but very spatially detailed at cities level [5]). Experiments have been led in two different contexts: first, the server was running on a standard dedicated machine, and latter, the computation tasks was parallelized and the server was unfolded on a Shared Multi-Processors (SMP) machine. In both cases, the client is connected via the secured SOAP protocol to the server. Since the server can be used independently of the client, and to avoid any alteration of measures introduced by net latencies, the performances were measured on the server side.

5.1   Experiments on a Standalone Server

Experiments have been made on the server's side running on a bi-processor machine, equipped with Linux OS (Pentium 4 at 2,6 Ghz, with 1 Gb of memory), which appears to be a very standard working machine. The complexity of the algorithm is ascertained by the computation time measures (see Table 1). This is linearly proportional to the resolution (total number n of points to be computed on the grid), to the number m of geographical entities (which is fixed in our experiments). The cost of the calculus is indirectly linked to the scope: a longer scope will involve more geographical units in the computation than a shorter one.

The complexity being known, this allows the user to estimate the duration of each map computation, according to the requested resolution and scope. This duration is quite reasonable when using standard computers: around 2 minutes for a long scope (100 km) and a fine resolution (800 x 600).

Table 1. Potential map computation duration

Resolution	Scope (km)	Calculation duration (s)
200x100	100	5
400x300	100	32
800x600	100	130
800 x 600	**25**	33
800 x 600	**50**	57
800 x 600	**100**	130

5.2   Experiments on a SMP Machine

Alternatively, the server can be unfolded on a SMP machine, shared by several users. Then the map is computed by pieces in a parallelized way by the set of requested processors.

The algorithm for the calculation of a potential map iterates on all points of the grid, which are arranged in an array of size n (n is the resolution level). The parallelization of computations does not present any difficulty due to interlaced dependances, since each point can be calculated independently of others. It is based on the distribution of tasks (a task is the calculation of an entry in the table) on a number of k processors. The number of points allocated to a processor is therefore n/k, and the calculation ends when each processor has finished calculating its portion of the table.

This naive distribution of tasks shows however two major limitations. Firstly, such an algorithm is not resistant to disturbance: if the capabilities of a processor are suddenly divided by 50%, the computing time is then extended by the time equivalent to half its task, since there is no process for automatically rebalancing workloads between processors. Secondly, our schedule is very heterogeneous: items located in areas with low density measurements are calculated much more quickly through the process of pruning, and thus some processors complete their work faster than others.

The mechanism of adaptive transmission load [15] is a solution to adapt dynamically the shared load: each "free" processor steals half of the remaining task to a busy processor. This redistribution occurs only when the busy processor completes its assigned indivisible portion of work α log (p). α is a configurable parameter, which is adjusted to reduce the duration time of the lock to access the data array: this time must be much shorter than the calculation duration of the portion of the table. This concept of minimum quantity of work is introduced to avoid problems of contention when accessing the array of points: when a processor takes a task, it activates a semaphore on the array and blocks access to other processors. This inter-blocking time is short, and the definition of a optimized minimum workload avoids the frequent repetition of inter-blockages.

This algorithm has been implemented and tested on a SMP, 8 cores with shared memory. The experiment gives results confirming the effectiveness of the method. Table 2 shows performances for a 800x600 resolution and a scope of 100 Km. Response time falls from 130 s for one processor to 32 s for 8 processors. In addition, we note that this algorithm has been implemented with two different libraries for scheduling tasks, TBB (Intel Threading Building Blocks, http://threadingbuildingblocks.org/) and Kaapi [4]. Whatever the library, performances are improved in the same way: the gain in computing time is linear with the number of processors mobilized, as long as the redistribution is not saturated by the memory bandwidth. Because of the amount of data transferred in our application, the saturation is reached from 5 processors with the pthread implementation, a bit further with Kaapi thanks to a memory-aware scheduling algorithm.

Table 2. Potential map computation duration for a resolution of 800x600 and scope of 100 km

Number of processors	1	2	3	4	5	6	7	8
Calculation duration (s)	130	104	72	50	41	40	36	32

6 Conclusion and Future Work

This paper presents HyperSmooth, the first system offering the possibility to interactively study the spatial propagation of social, environmental and economic phenomena through the potential transformation method. HyperSmooth is based on a client-server architecture. The Web-based client makes possible the visualization of a set of interactive maps through a user-friendly user-interface, and drives a server which efficiently computes maps (this computation can be distributed on a grid). The client-server communication relies on the SOAP protocol and the security of data is granted by SSL cryptography. Interactivity requires a short time response from the system to the user's queries. For that purpose, we have developed *ad-hoc* optimization techniques, which shorten the computation duration of potential maps for the server. Calculations are speeded-up by an algebraic cut-off method and tabulation of orthodromic distances. Experiments on real dataset show good performances. In addition, the server side named *Hyantes* has been developped under an open-source licence, and can be downloaded freely on its Web site[1], allowing users to exploit and extend it for further experiments.

Currently, we are working on improving calculation performances by replacing the orthodromic distance with the euclidian one for small regions, and adopting a strategy for sub-sampling data according to the observation scale. We are also enhancing the client's functionalities by triggering potential map computation using query windows, and introducing iso-potential curves to overlay on the vectorial background map. Finally we think to couple our system with a Spatial OLAP tool in order to take advantage from its scalability, hierarchical and multidimensional data management features and usability.

References

1. Andrienko, G., Andrienko, N., Gitis, V.: Interactive Maps for Visual Exploration of Grid and Vector Geodata ISPRS. Journal of Photogrammetry and Remote Sensing 57(5–6), 380–389 (2003)
2. Bimonte, S., Wehrle, P., Tchounikine, A., Miquel, M.: GeWOlap: a Web Based Spatial OLAP Proposal. In: Second International Workshop on Semantic-based Geographical Information Systems, Montpellier, France, October 29-30. Springer, Heidelberg (2006)
3. Compieta, P., Di Martino, S., Bertolotto, M., Ferrucci, F., Kechadi, T.: Exploratory spatio-temporal data mining and visualization. Journal of Visual Languages and Computing 18(3) (2007)
4. Danjean, V., Gillard, R., Guelton, S., Roch, J.-L., Roche, T.: Adaptive Loops with Kaapi on Multicore and Grid: Applications in Symmetric Cryptography. In: Parallel Symbolic Computation (PASC 2007), London, Ontario. ACM publishing, New York (2007)
5. Dubois, A., Gensel, J., Hanell, T., Schürmann, C., Lambert, N., Zanin, C., Ysebaert, R., Grasland, C., Damsgaard, O., Lähteenmäki-Smith, K., Gloersen, E., Thomas, R.: Observing the structure of European territory in relative terms DG-IPOL, Regional Disparities and Cohesion: What Strategies for the future, EU Parliament, report IP/B/REGI/IC/2006_201 (2007)
6. Edsall, R., Andrienko, G., Andrienko, N., Buttenfield, B.: Interactive Maps for Exploring Spatial Data. In: Madden, M. (ed.) ASPRS Manual of GIS (2008)
7. Keenan, P.: Using a GIS as a DSS Generator. In: John, D., Jenny, D., Thomas, S., et al. (eds.) Perspectives on Decision Support System, pp. 33–40. University of the Aegean, Grèce (1996)
8. Kraak, M.-J., Brown, A. (eds.): Web Cartography. Taylor&Francis, London (2000)

---

[1] http://hyantes.gforge.inria.fr/

9. Gotway, C., Young, L.: Combining Incompatible Spatial Data. Journal of the American Statistical Association 97(458), 632–648 (2002)
10. Grasland, C., Mathian, H., Vincent, J.-M.: Multiscalar Analysis and map generalisation of discrete social phenomena: Statistical problems and political consequences. Statistical Journal of the United Nations ECE 17, 1–32 (2000)
11. Josselin, D., Fabrikant, S.: Cartographie animée et interactive, Revue Internationale de Géomatique, Editions Hermès (2003)
12. Longley, P., Goodchild, M., Mahuire, D., Rhind, D.: Geographic Information Systems and Science, p. 517. John Wiley & Sons, New York (2001)
13. MacEachren, A., Kraak, M.-J.: Exploratory cartographic visualization: advancing the agenda. Computers & Geosciences 23(4), 335–343 (1997)
14. Openshaw, S., Taylor, P.J.: A million or so correlation coefficients Statistical methods in the spatial sciences, Pion, London, pp. 127–144 (1979)
15. Roch, J.-L., Traore, D., Bernard, J.: Processor-oblivious parallel stream computations. In: 16th Euromicro International Conference on Parallel, Distributed and network-based Processing, Toulouse, France (2008)
16. Tobler, W.: Frame independant spatial analysis. In: Goodchild, M., Gopal, S. (eds.) The accuracy of spatial databases, NCGIA, ch. 11, pp. 115–119. Taylor & Francis, Abington (1991)
17. W3C. SOAP., http://www.w3.org/TR/soap/

SVG-Based Spatial Information Representation and Analysis

Haosheng Huang[1,*], Yan Li[2], and Georg Gartner[1]

[1] Institute of Geoinformation and Cartography, Vienna University of Technology, Austria
[2] School of Computer, South China Normal University, China
huanghaosheng@gmail.com, yanli@scnu.edu.cn,
georg.gartner@tuwien.ac.at

Abstract. This paper tries to make some spatial extensions to W3C's Scalable Vector Graphics (SVG) Specification to support SVG-based spatial information representation and analysis in the Web environment. Based on spatial data modeling, this paper attempts to find a theoretical foundation for SVG-based spatial information representation. Then we propose an SVG-based spatial information representation model for spatial data publishing. Furthermore, this paper designs and implements some spatial operators, and integrates them into an SVG-based Spatial Extended SQL to support spatial analysis, which improves the functions of current WebGIS applications, most of which have been only employed for visualization. Finally, this paper designs some case studies. The results of the case studies prove that the suggested methods are feasible and operable for spatial information publishing and analysis via Web.

Keywords: SVG-based WebGIS, spatial data modeling, spatial information representation, Web-based spatial analysis, SVG-based Spatial Extended SQL.

1 Introduction

With the rapid development of the Internet/Web, WebGIS is an inevitable trend. WebGIS aims at providing GIS functions (such as web mapping, spatial analysis) to users via Web environment.

SVG (Scalable Vector Graphics) was introduced by W3C in 2001, since then, it has become an active area of research in WebGIS. Some researchers use SVG's shape elements (such as *line*, *path*, etc.) and graphic styles (such as *fill*, *stroke*, etc.) for spatial information visualization and develop some prototype systems [1] and [2]. Furthermore, there are also some researches focusing on transformation of GML to SVG [3] and [4]. As SVG is developed primarily as a publishing tool of 2D graphics, however spatial information has its own characteristics in representing and organizing spatial objects and their relationships (such as hierarchical structure of map - layer - spatial object, spatial attributes vs. non-spatial attributes). In order to use SVG for spatial information

---

* Corresponding author.

M. Bertolotto, C. Ray, and X. Li (Eds.): W2GIS 2008, LNCS 5373, pp. 17–26, 2008.

publishing, a model which takes the characteristics of spatial information into account has to be developed. Unfortunately, little work has been done on this topic.

At the same time, currently, lots of SVG-based WebGIS applications have been designed for visualization only, but avoid the access to spatial analysis functions such as spatial topological query, map overlay, and buffer that are vital for spatial applications [5]. This is mainly due to the fact that there are no effective functions to support spatial query and analysis directly on SVG.

This paper attempts to solve the two problems mentioned above. First, based on the theory of spatial data modeling, this paper tries to find a theoretical foundation for SVG-based spatial information representation. And then, based on this, we propose an SVG-based spatial information representation model for publishing spatial data. In order to support SVG-based spatial analysis, we design some spatial operators, and integrate them into SVG-based Spatial Extended SQL. With these suggested methods, users can easily carry out spatial information publishing and analysis tasks via Web.

The paper is arranged as follows. Section 2 describes an SVG-based spatial information representation model. In section 3, we discuss SVG-based spatial analysis. Some case studies are implemented to evaluate our suggested methods in section 4. Finally, the conclusions and future work will be described in section 5.

2   SVG-Based Spatial Information Representation

2.1   Theory of Spatial Data Modeling

Spatial data modeling is the process of abstracting the real world (identifying the relevant objects and phenomena) and representing it in an appropriate form which can be recognized by computers [6]. As SVG is one of the forms which computers use to represent relevant objects or phenomena of the real world, thus we can utilize the theory of spatial data modeling, to discuss SVG-based spatial information representation which attempts to represent the relevant objects and phenomena.

There are three models of spatial data modeling: conceptual model, logical model, physical model. According to the contents and requirements of these models, spatial data modeling includes three steps [6]: 1) choose a conceptual model which can abstract the real world most appropriately, 2) choose an appropriate data structure to represent the conceptual model, 3) design a file format, or the appropriate method of record or storage for the data structure in step 2.

In the following sections, we use the above mentioned steps to develop an SVG-based spatial information representation model. First, we make some extensions to the Open Geospatial Consortium (OGC)'s Geometry Object Model [7], and take the extended model as our conceptual model. And then we design a spatial data structure for this conceptual model based on object-oriented design. Finally, we use SVG standard (file format) to represent the above spatial data structure.

2.2   Spatial Conceptual Data Model

Spatial conceptual data models can be categorized into raster data model and vector data model. The latter treats the world as surface littered with recognizable spatial objects (e.g. cities, rivers), which exist independently of their locations [8]. As SVG is developed to represent vector graphic, we will focus on vector data model.

In vector data model, spatial data is organized as a hierarchical structure: *spatial entity (object)*, *layer* and *map*. *Spatial entity* refers to the thing or phenomenon which has geometrical shape. It has two types of attributes: *spatial attributes* that describe geometry and topology of the spatial entity, and *non-spatial attributes* which define the semantics (name, theme, etc.) of the spatial entity. Spatial entities belonging to the same theme always have similar non-spatial attributes or geometrical type, and thus are grouped into a *layer*. A *map* is always constituted by different layers which describe the same region of the real world.

OGC's Geometry Object Model (GOM) [7] is one of the most popular vector models. It abstracts spatial entities as *Point, Curve, Surface, Multipoint, Multicurve* and *Multisurface* according to their geometries. *Line* is a special type of *Curve*, and *Polygon* is a special type of *Surface*. As GOM only defines spatial attributes of spatial entities, we extend it by adding some concepts like map, layer and non-spatial attributes. Also we don't depict the relationships between *Point* and *Curve*, *Curve* and *Surface*. The reason is that we want to have a relative simple SVG structure (without too many *"xlink:href"*), which will improve the rendering and querying performance of SVG document. Fig.1 shows our conceptual model.

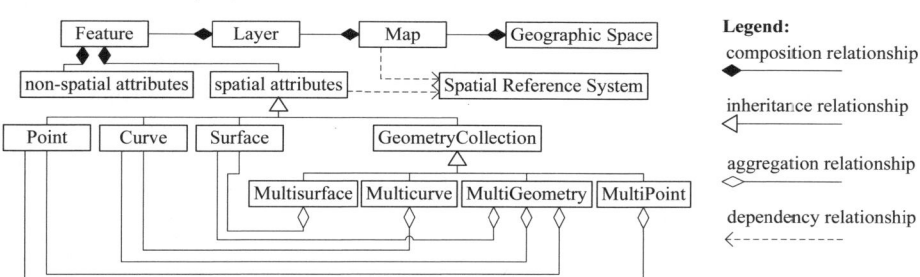

Fig. 1. The conceptual model (using UML notations) which is extended from GOM (after [7])

2.3 Spatial Data Structure

In this section, we design a spatial data structure for the above conceptual model based on object-oriented design. Fig. 2 depicts the spatial data structure.

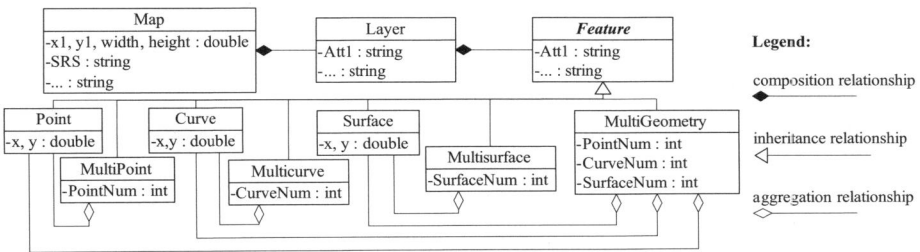

Fig. 2. The class hierarchy of spatial data structure which represents the conceptual model

In this data structure, spatial entity is designed as the abstract class *Feature*. The non-spatial attributes of spatial entity are designed as data members of the class *Feature*. *Point, Curve, Surface, Multipoint, Multicurve, Multisurface* and *Multigeometry* are inherited from the class *Feature*. Map is described as the class *Map*, which includes data members such as *x1, y1, width, height* (the bounding of the map) and *SRS* (Spatial Reference System).

2.4   SVG-Based Spatial Information Representation Model (File Format)

In this section, we discuss how to use SVG standard to represent (store) the above spatial data structure. The basic principles are as follows: 1) Use *<svg>* element to represent the class *Map*, and use *viewBox* attribute to represent data members *x1, y1, width, height*; 2) The abstract class (*Feature*) doesn't need to be represented; its data members are represented in its inherited classes; 3) Data members (spatial attributes and non-spatial attributes) in a class are represented as corresponding SVG element's attributes; 4) If class B is PART-OF class A (composition/aggregation relationship in the spatial data structure), use *<g>* element to represent the class A and group its members (e.g., class B.) together. For example, the class *Feature* is PART-OF the class *Layer*, so we use *<g>* element to represent the class *Layer* and group the class *Feature* (spatial entity) together. According to these principles, *<g>* element is used to represent the class *Layer, Multipoint, Multicurve, Multisurface* and *Multigeometry*.

Fig. 3 depicts the SVG-based spatial information representation model.

In this model, if B is PART-OF A, B will be represented as a child element of B. For example, *Layer* is PART-OF *Map*, so *<g>* element which represents *Layer* is a child

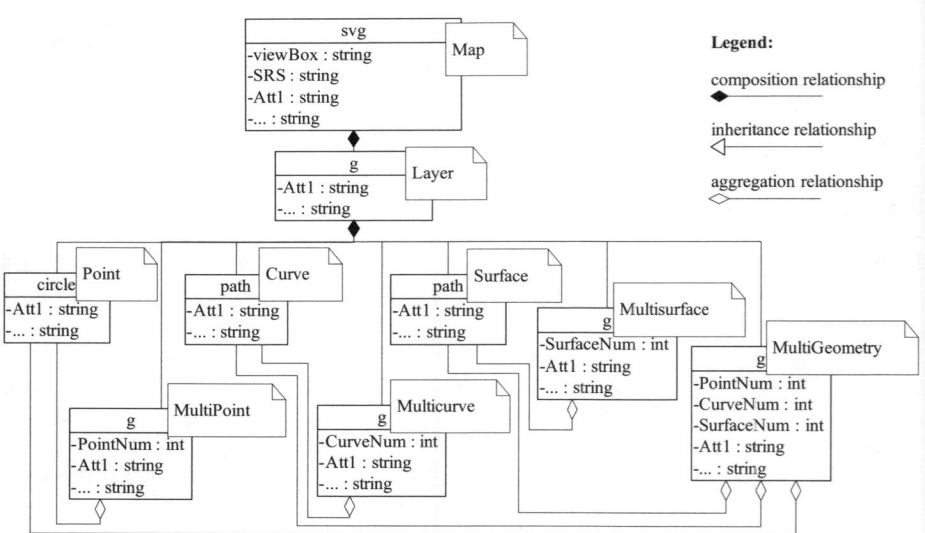

Fig. 3. SVG-based spatial information representation model

element of *<svg>* element which represents the *Map*. With this model, users can use SVG to publish spatial information. Please refer to section 3.2 for an example.

3 SVG-Based Spatial Analysis

The ability of spatial analysis is viewed as one of the key characters which distinguish GIS from CAD systems or other information systems. Currently, SVG-based WebGIS has been employed for visualization only; but avoided the access to spatial analysis functions. This is mainly due to the fact that there are no effective web-based functions to support spatial query, and analysis on SVG.

Since the need for spatial query languages has been identified in 1982 [9], many authors have designed their spatial extended query language [5] and [10]. These SQL like languages introduce spatial data types (e.g., point, line and polygon) and spatial operators, allow users to inquire spatial features, primarily in terms of spatial relationships and metric constraints [5]. It is widely acknowledged that these spatial operators and SQL like languages can be used for spatial analysis.

According to section 2, spatial information is organized as map - layer - spatial object. A map represented by SVG can be viewed as a database, the layers as tables of the database, the attributes (spatial and non-spatial) of spatial objects in the layer as columns of corresponding table, spatial objects as records of corresponding table. Thus SQL like languages can be used for spatial analysis on SVG.

In this section, we design some spatial operators and integrate them into our SVG-based Spatial Extended SQL (SSESQL). This SSESQL uses the basic spatial data types discussed in section 2: *Point, Curve, Surface, Multipoint, Multicurve, Multisurface* and *Multigeometry*. The proposed SSESQL can be used on server side or browser side for spatial query and analysis on SVG.

3.1 Spatial Operators

Spatial operators are mainly designed to access spatial attributes, calculate spatial relationships, and perform geometrical operations. We design the following operators:

1) Attribute access operators: They include *GeometryType, Centroid, Length, Area,* and *Envelope*. They are used to calculate length, area and centroid of spatial object.

2) Spatial topological operators: Spatial topological relationship is very important for spatial query and spatial analysis. There are two different approaches describing the spatial topological relationship: DE-9IM and RCC-8. According to [11], these two completely different approaches lead to exactly the same set of topological relations. This paper uses the smallest complete set of topological relationship based on DE-9IM: *Disjoint, Touch, Crosses, Within* and *Overlap*. Thus, spatial topological operators include *Disjoint, Touch, Crosses, Within* and *Overlap*. For convenience, *Contain* operator is also included as the opposition of *Within* operator.

3) Spatial order operators: This kind of operators includes: *East, East_South, South, West_South, West, West_North, North* and *East_North*.

4) Spatial metric operators include *Max_Dist, Min_Dist* and *Distance* operators.

5) Geometrical operators: Sometimes, spatial analysis needs to create new spatial features with some geometrical operations. As proved in mathematics, *{Intersection, Union, Difference}* is a complete set for 2D geometrical operations. So we define *Intersection*, *Union*, and *Difference* as geometrical operators. In order to support Buffer analysis, *Buffer* operator is also designed for this kind of operators.

These five kinds of operators can meet the basic requirements of spatial analysis. For network analysis, we can use *Touch* operator and *Length* operator to find out the touched spatial object (e.g., road) and the distance. We can use *Buffer* operator and topological operators to carry out buffer analysis. Also, we can use *Intersection, Union* and *Difference* operators to carry out overlay analysis.

3.2 SSESQL and Some Query Examples

As SSESQL is designed for spatial query, there is no need to consider data insert, update and delete. All we should do is integrating the above spatial operators to the original SELECT clause of SQL. Please refer to [12] for the EBNF (Extended Backus-Naur Form) of SELECT clause of SSESQL.

The following SVG codes show a map of Guangdong province in China which is based on our suggested model (Fig. 3). This map includes two layers: city and river.

```
<svg viewBox="94928 2172873 790615 595213" SRS="xi'an80">

  <g id="city">

    <path id="C1" pop="1500000" d="…"/>

    …

  </g>

  <g id="river">

    <path id="R1" length="100" d="…"/>

    …

  </g>

</svg>
```

The two layers expressed in the above SVG can be viewed as the following tables: city (id, pop, d) and river (id, length, d). The following are some query examples.

1) Query example 1: List the cities which are crossed by the river "R1".

```
SELECT c.id AS cid  FROM river r, city c

WHERE r.id="R1" AND Crosses(r.d, c.d)=True;
```

2) Query example 2: River "R1" can supply water for the cities, which are 40KM around the river. List this kind of cities.

```
SELECT cy.id  FROM city cy, river r

WHERE Overlap(cy.d, Buffer(r.d, 40))=True AND r.id="R1";
```

3.3 SSESQL-Based Spatial Analysis in the Web Environment

The general workflow of spatial analysis includes four steps [13]: 1) define the goal and evaluation criteria; 2) prepare and represent the needed spatial data; 3) carry out spatial query and analysis with GIS tools; 4) result appraisal and explanation. Step1 and step 4 need domain knowledge and are mainly carried out by experts. Step 2 and step 3 need GIS tools to support human-computer interaction.

For step 2, we can use SVG to represent the needed spatial data based on the representation model depicted in Fig. 3; while for step 3, SSESQL can be utilized to carry out spatial query. Fig. 4 shows the workflow of SVG-based spatial analysis.

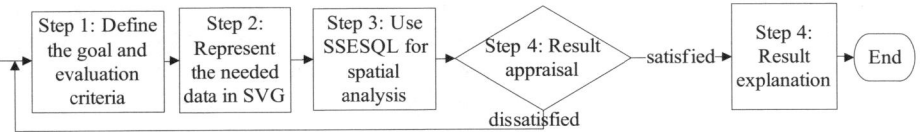

Fig. 4. Workflow of SVG-based spatial analysis

With the above workflow, users can easily use SVG to carry out spatial analysis task in the Web environment.

4 Implementation and Case Studies

4.1 Implementation of Spatial Operators and SSESQL

It is widely acknowledged that load-balancing between server side and browser side is a feasible technology for Web-based spatial analysis. WebGIS will carry out spatial operations on server side or browser side based on communication cost versus computational cost. As a result, we need to develop the spatial operators and SSESQL for both server side and browser side.

Algorithms of *computational geometry* can be utilized to develop our spatial operators. As Java has provided some basic computational geometry APIs (Java 2D API), we implement the spatial operators with algorithms of *computational geometry* and Java 2D API. For the SSESQL, we develop a compiler to carry out syntax, sentence, and semantic analysis for SSESQL sentences. For the server side implementation, spatial operators and SSESQL compiler are designed as Java servlets. For browser side implementation, spatial operators and SSESQL compiler are developed as Java applets, and embedded in HTML; JavaScript DOM APIs of HTML and SVG are used to access SVG document and invoke the Java applets; a user interface is also embedded in HTML for inputting SSESQL query sentences (Fig. 5). Users can access spatial analysis functions simply with a web browser (such as, IE) which contains an SVG plug-in or SVG viewer (such as Adobe SVG Viewer).

In the following, we design two case studies to evaluate the suggested methods. As load-balancing between server side and browser side is not this paper's research focus, we carry out spatial analysis based on the browser side implementation.

4.2 Case Studies

The cultivated lands are very important in a highly populated area like Guangdong Province in China. People may consider whether the cultivated lands change with the growth of transportation network. We choose two issues to analyze and discuss.

The first issue is how the cultivated lands along the railway and highway change. Let's take 15km extent and analyze the changes of cultivated land between 1987 and 1999. We carry out this task based on the method described in Fig. 4. First, based on the suggested model in Fig. 3, we use SVG to represent the needed spatial data (railway, highway, district map, and different years' statistics on land use). And then use SSESQL to carry out spatial queries on this SVG. Fig. 5 shows the user interface and result of this case study. Functions of zoom in/out, roam, layer control, query, and statistics are also added to this interface. This task uses the *"Buffer"*, *"Union"*, *"Within"* operators and SSESQL in the analyzing procedure. The right low box lists the names of cities involved in the calculation. At last, we use the statistics function to generate the bar graphs of changes of land use for every relevant city.

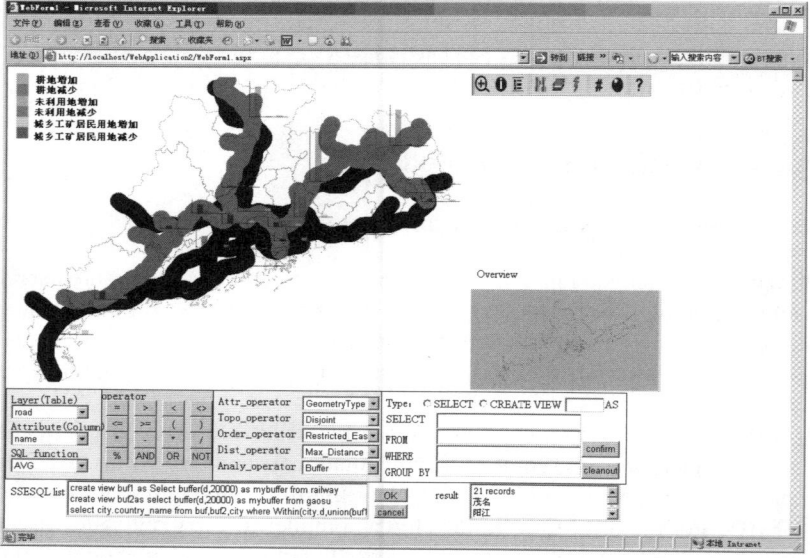

Fig. 5. The cultivated lands change along the railway and main highway. (Legend for bar graphs: dark orange for cultivated lands increasing; dark green for cultivated lands decreasing; middle orange for unused land increasing; middle green for unused land decreasing; light orange for residential area increasing; dark grey for residential area decreasing).

Another issue is: what's the relationship between cultivated lands and road density (railway, highway, and province-level road). We use the *"Interaction"*, *"length'*, *"area"* operators and SSESQL to carry out this task step by step. First, we utilize the *"intersection"* and *"length"* operators to calculate the total road length for every city, and use *"area"* operator to calculate every city's area, and then calculate every city's road density. Fig. 6 depicts the result of this issue. The result lists the name of cities and their road density at the right low box. Every city is marked with different color according to its road density, and with a statistic bar graph of changes of land use. The legend of the bar graph is the same as that in Fig. 5.

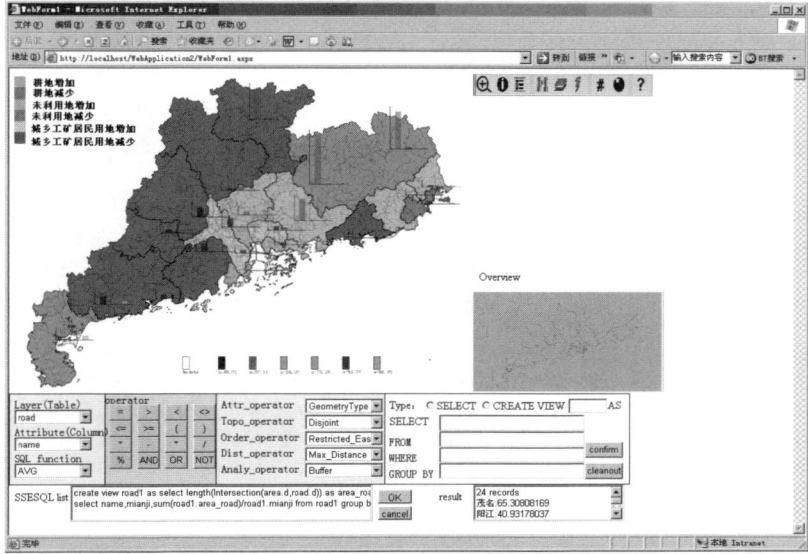

Fig. 6. The relationship between cultivated lands and road density

The implementations of the above case studies show that our suggested methods for SVG-based publishing and analysis are feasible and operable. The SSESQL can also be extended to carry out spatial query on GML, which can be used to provide load-balancing spatial analysis for XML/GML/SVG based WebGIS.

5 Conclusions and Future Work

This paper tries to make some spatial extensions to W3C's SVG Specification to support SVG-based spatial information representation and analysis in the Web environment. Based on the theory of spatial data modeling, this paper sets up a theoretical foundation for SVG-based spatial information representation, and then develops an SVG-based spatial information representation model based on this theoretical foundation. Furthermore, we design and implement some spatial operators, and integrate them into our SVG-based Spatial Extended SQL (SSESQL), and then discuss how to use SSESQL for spatial analysis. In order to evaluate our suggested methods, this paper designs and implements two case studies. The results of the case studies show that the proposed methods are feasible and operable to support spatial information publishing and analysis in the Web environment.

Our next step is to consider the load-balancing technology between server side and browser side for SVG-based spatial analysis. We are also interested in the problem of completeness of the spatial operators for spatial analysis task.

Acknowledgments. This work is supported by SemWay project (by Austrian BMVIT), UCPNavi project (by Austrian FWF), and Guangdong's project (2005B30801006, by Science and Technology Department). We are also grateful to our anonymous reviewers for their truly helpful comments.

References

1. Cartographers on the net, http://www.carto.net
2. Li, Q., Xie, Z., Zuo, X., Wang, C.: The spatial information description and visualization based on SVG (in Chinese). Acta geodaetica et cartographica sinica 34(1), 58–63 (2005)
3. Guo, Z., Zhou, S., Xu, Z., Zhou, A.: G2ST: A novel method to transform GML to SVG. In: Proceedings of the 11th ACM GIS, pp. 161–168 (2003)
4. Tennakoon, W.T.M.S.B.: Visualization of GML data using XSLT. Diploma Thesis, International Institute for Geo-Information Science and Earth Observation (2003)
5. Lin, H., Huang, B.: SQL/SDA: A query language for supporting spatial data analysis and its Web-based Implementation. IEEE TKDE 13(4), 671–682 (2001)
6. Chen, S., et al.: Introduction of GIS (in Chinese), pp. 28–30. Science publish, Beijing (2001)
7. Open GIS simple feature specification for SQL,
   http://www.opengeospatial.org/standards/sfs
8. Shekhar, S., Coyle, M., Goyal, B., Liu, D., Sarkar, S.: Data models in geographic information systems. Communications of the ACM 40(4), 103–111 (1997)
9. Frank, A.: Mapquery-database query languages for retrieval of geometric data and its graphical representation. ACM Computer Graphics 16(3), 199–207 (1982)
10. Egenhofer, M.: Spatial SQL: A Query and Presentation Language. IEEE Transactions on Knowledge and Data Engineering (TKDE) 6(1), 86–95 (1994)
11. Renz, J., Rauh, R., Knauff, M.: Towards Cognitive Adequacy of Topological Spatial Relations. In: Habel, C., Brauer, W., Freksa, C., Wender, K.F. (eds.) Spatial Cognition 2000. LNCS, vol. 1849, p. 184. Springer, Heidelberg (2000)
12. Huang, H.: Key issues of SVG-based Network spatial Information Publishing (in Chinese), Master Thesis, South China Normal University (2006)
13. Wu, X.: Principles and methods of GIS (in Chinese), pp. 156–157. Publishing House of Electronics Industry, Beijing (2002)

Design and Implementation of GeoBrain Online Analysis System (GeOnAS)

Weiguo Han, Liping Di, Peisheng Zhao, Yaxing Wei, and Xiaoyan Li

Center for Spatial Information Science and System, George Mason University,
6301 Ivy Lane, Suite 620, Greenbelt, MD 20770, USA
{whan,ldi,pzhao,ywei,xlia}@gmu.edu

Abstract. GeOnAS is an extensible, scalable and powerful online geospatial analysis system based on Service Oriented Architecture (SOA), and is designed and implemented with the complementary technologies, Asynchronous JavaScript and XML (Ajax) and Web services, which greatly increase the interactive capabilities of graphical user interfaces and improve the user experience. It provides a highly interoperable way of accessing Open Geospatial Consortium (OGC) compliant web services for geospatial data discovery, retrieval, visualization and analysis, and leverages web service standards to enable service discovery, selection, negotiation and invocation for making more informed decisions.

Keywords: Web Service, Service Oriented Architecture, Asynchronous JavaScript and XML, Open Geospatial Consortium Specification, Online Analysis.

1 Introduction

Today many web-based geospatial applications are being built to help Geosciences researchers solve real-world problems. However, only a few could help scientists easily combine multiple, disparate geospatial datasets and provide powerful and advanced analysis functionality that goes beyond geospatial data visualizations.

A web service-oriented online geospatial analysis system, named GeOnAS (http://geobrain.laits.gmu.edu:81/OnAS/), has been implemented to make petabytes of data and information from NASA's Earth Observing System (EOS) and other earth science data providers easily accessible, and provide value-added geospatial service and modeling capabilities to geosciences community with a standard web browser and internet connection [1].

This paper describes design and implementation of GeOnAS. The reminder of this paper is organized as follows: section 2 reviews the previous and current progresses in this field; section 3 presents the system general architecture and data flow briefly; in section 4, each module of GeOnAS is described in details; finally, section 5 concludes and points to future work.

2 Overview

Geosciences researchers opt to use desktop geospatial software packages such as ArcInfo or PCI because they are more interactive and responsive than Web applications. But they

M. Bertolotto, C. Ray, and X. Li (Eds.): W2GIS 2008, LNCS 5373, pp. 27–36, 2008.
© Springer-Verlag Berlin Heidelberg 2008

must purchase the commercial software, obtain geospatial datasets from various sources, preprocess these datasets, and analyze the datasets on their local machines [1].

Interactive Visualizer and Image Classifier for Satellites (IVICS), a free download-able desktop visualization tool to facilitate selection of training samples from satellite images, has evolved into a general purpose visualization system that supports several common satellite and remote sensing data formats [2]. Multiple-Protocol Geospatial Client (MPGC), the predecessor of GeOnAS, provides an interoperable way of ac-cessing OGC-compatible geospatial Web services for integrating and analyzing dis-tributed heterogeneous Earth science data [3]. However, users must create the proper running environment for these tools on their computers.

Global Land Cover Facility at University of Maryland developed the Earth Science Data Interface for searching, browsing and downloading data [4], and Path/Row Search, Map Search and Product Search are provided to meet the requirements of the international scientific and educational communities [5]. However, geospatial data customization is not provided for user's convenience.

Earth Science Gateway (ESG) of NASA streamlines access to remote geospatial data, imagery, models, and visualizations through open, standard web protocols. By organizing detailed metadata about online resources into a flexible, searchable regis-try, ESG lets scientists, decision-makers, and others access a wide variety of observa-tions and predictions of natural and human phenomena related to Earth Science [6]. ESG provides an easy, efficient and useful approach to integrate various geospatial systems and components through open interfaces [7]. Nevertheless, its capabilities for online geospatial analysis should be enriched.

Service Oriented Architecture (SOA) offers a fresh and flexible approach and an adaptive and responsive architecture for the development of web geospatial applica-tions. Web services and Ajax enable developers to leverage web application in new extendable ways. These technologies are utilized in the implementation of GeOnAS.

3 Architecture

Figure 1 depicts the general architecture of GeOnAS. The fundamental design princi-ple is to leverage distributed computational resources to support geospatial analysis and decision making [1].

The rich-browser client in the architecture is a web browser with Document Object Model (DOM), Dynamic HTML (DHTML), Extensible Markup Language (XML), Cascading Style Sheets (CSS), and JavaScript. This client allows processing of all presentation logics and presents all user functions via the standard web browser (i.e. Internet Explorer or Firefox). Intuitive graphical user interfaces, including pull down menus, trees list, toolbar, modal/modeless popup dialogs, and tabbed dialogs are im-plemented like those of desktop software bases on AJAX. Especially, user does not need to refresh the full page because XMLHttpRequest object in JavaScript is utilized to decouple user interactions in the browser from the browser requests to remote services [8].

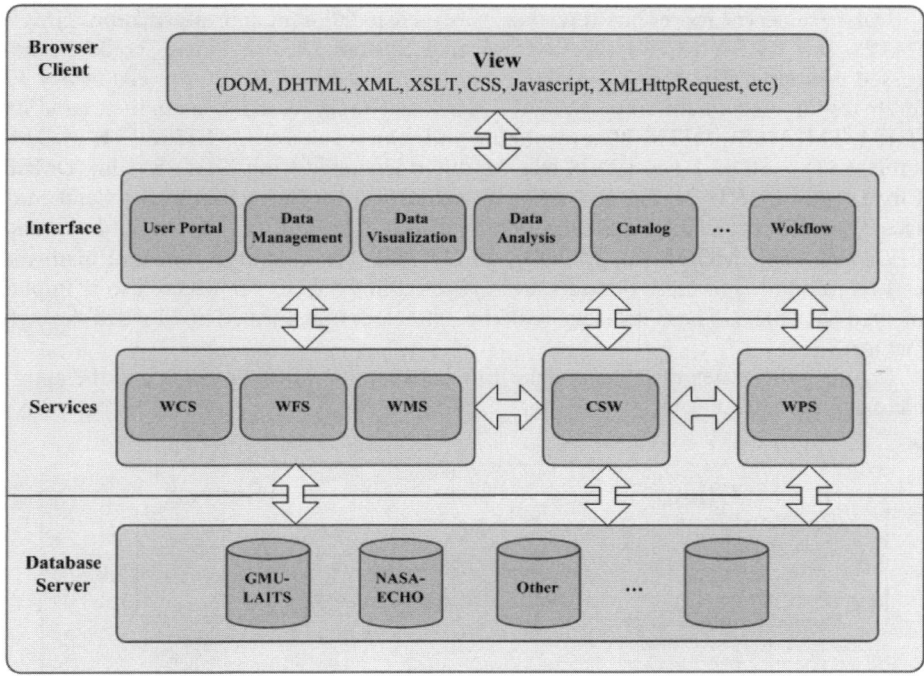

Fig. 1. General Architecture

Interface includes the models of User Portal, Data Management, Data Visualization, Data Analysis, Catalog, Workflow, etc. User Portal enables user to store the portal state in a WMC file for future use by including a list of related data layers and processing services. Data management provides user with geospatial data retrieval via a network accessible point. Data Visualization provides a set of different rendering services to build their own preference. Catalog is targeted at metadata of geospatial data and processing services. Data Analysis offers the powerful processing services to perform advanced geospatial analysis. Workflow allows user to build a chain of services to perform the real world task [1]. These functions are implemented as Servlets and JavaServer Pages (JSP).

Services in the framework represent OGC specification compliant services. Along with Web Coverage Service (WCS)[1] developed by Center for Spatial Information Science and System (CSISS), George Mason University, the framework provides extensibility of other OGC standards, such as Catalogue Services for Web (CSW), Geographic Markup Language (GML), Web Map Context (WMC), Web Feature Service (WFS), Web Map Service (WMS), and Web Processing Service (WPS). Support for these standards enables GeOnAS to interoperate successfully and conveniently with other vendor web services across the globe.

---

[1] http://data.laits.gmu.edu:8080/pli/www/wcs110.htm

Database server represents repositories of geospatial data and information. GMU-LAITS and NASA ECHO (Earth Observing System Clearinghouse) could be accessed currently. The former one built and maintained by CSISS provides nearly 12 terabytes of raster data from NASA, USGS and other sources, including LandSat (ETM, TM, MSS), SRTM(30m resolution and 90m resolution), ASTER(L1B in Geo-TIFF, L1B in HDF, L2 in HDF), BlueMarble (Generated from MODIS data), DMSP City Lights, EO1(Hyperion and ALI), WindSat(Soil Moisture Retrievals, Land Surface Temperature, Land Type, Observation Time), NetCDF(NOAA GOES data), and CEOP(MOD05, MOD11L2, MYD05, MYD11L2) for customization and analysis, and metadata tied to these datasets are registered in CSW server which also is implemented by CSISS. These datasets could be extracted, transformed and loaded through GeOnAS.

Figure 2 illustrates the typical data flow between the browser client and the application server in GeOnAS.

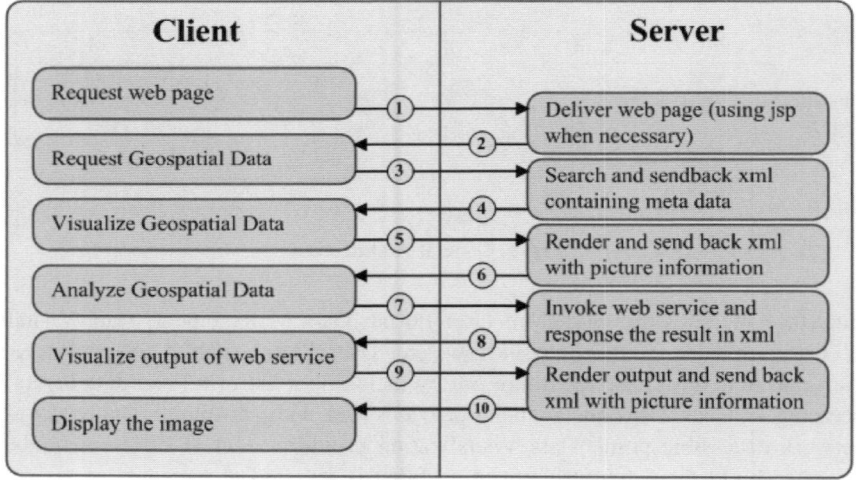

Fig. 2. Data Flow

4   Implementation

GeOnAS is deployed in an Apple G5 Cluster Server, which is composed of 1 Head Node, 4 Cluster Nodes, 3 Xserver RAIDs, 2Gbps Fiber Switch, 8-Port Gigabit Ethernet Switch, and other hardware [9]. The running environment could provide worldwide users with online heterogeneous geospatial data access and multi-source and powerful geospatial services at a high availability.

Now, GeOnAS includes the modules of Project Management, Data Manipulation, Map Operation and Display, Vector and Raster Data Analysis, Web Service Invocation and Auxiliary functions.

4.1 Project Management

This module provides Create Project, Open Project, Save Project, Close Project, and Exit functions.

When creating a new project, user could specify the bounding box (by location name, inputting coordinates, or dragging box on the map) and coordinate reference system (CRS) of the project through an improved Google Map interface.

The project file could be saved as a WMC file to local disk for future use anywhere or data sharing, or as a Keyhole Markup Language (KML) file to integrate into Google Earth or Google Map. Figure 3 shows the false color composition output, which is shown in Figure 5 below, in Google Earth.

Fig. 3. False Color Composition Output Display in Google Earth

4.2 Data Manipulation

This module contains Data Query, Data Selection and Customization, Data Adding and Removal, Data Rendering, Data Export, and so on.

User could specify query constraints, including temporal information, keywords, sensor name, platform name, instrument name etc, to perform query operations.

As shown in Figure 4, based on the query result set, the user interface of dataset selection provides preview, size, format, and description, etc. metadata information of geospatial dataset which could help user determine whether the datasets meet their requirements.

The selected datasets could be added with the customized resolution or width and height, and be rendered with the preferred palette in GeOnAS.

Datasets in the project could be exported to the local disk in multiple common formats (such as GeoTIFF, HDF) and specified projection coordinates via Data Export function, so user could make use of these exported file in other geospatial software.

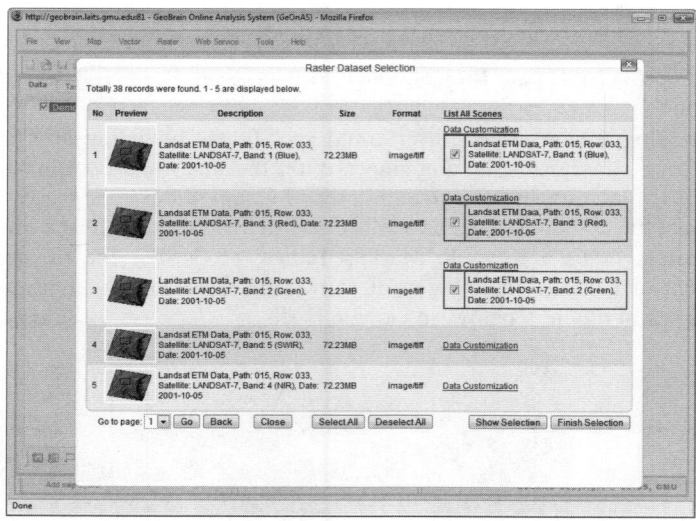

Fig. 4. Dataset Selection

4.3 Map Operation and Display

Zoom In/Out, Pan, Full Extent, Move Forward/Backward, Refresh, Overview and other basic map commands are provided to be similar to those of the common GIS and RS desktop software.

This module also allows user to show or hide map layer, control display order, export image, and set palette. For example, when a layer is selected and moved up or down in the layer tree, its display order will change on-the-fly.

4.4 Vector and Raster Data Analysis

Vector data analysis contains Feature Query, Feature Extraction, and Shortest Path function through invoking the corresponding web services registered in the CSISS CSW server.

Raster data analysis includes web NDVI, Image Algebra, Image Slice, Color Transformation (RGB2HIS, HIS2RGB), RGB Composition (True Color, False Color), RGB Extraction, Clipping, Image Patch, Image Mosaic, Classification (Supervised, Unsupervised) services.

These services wrap many of the associated functions of the open source GIS software Geographic Resources Analysis Support System (GRASS) [10], and are registered in CSISS CSW server so that they could be managed, discovered, and used effectively. The detailed description, including operations, input and output parameters, service location, Web Services Description Language (WSDL), etc, could be got at http://geobrain.laits.gmu.edu:81/grassweb/manuals/index.html. All these services are self contained and only loosely coupled with each other, so they could be composed into work flow to create a composite application.

Figure 5 illustrates the false color composition in GeOnAS.

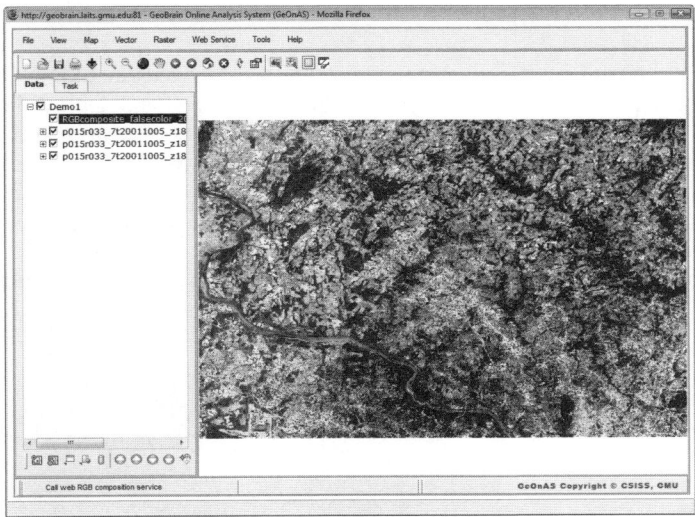

Fig. 5. False Color Composition

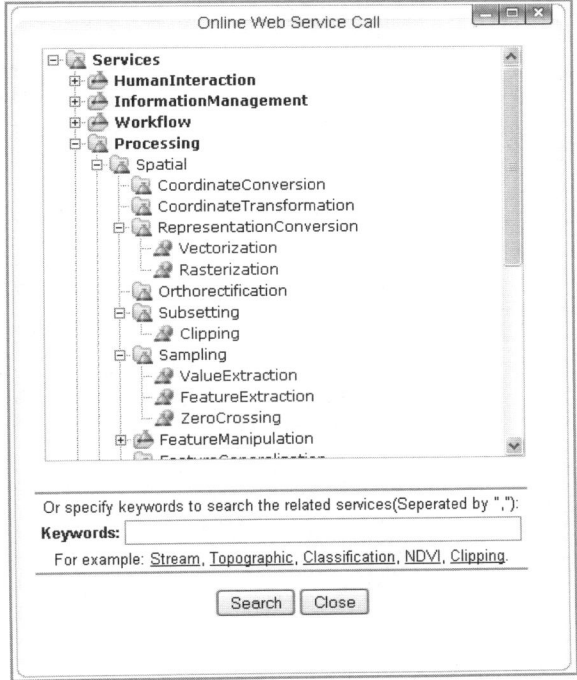

Fig. 6. Web Services Tree List

4.5 Web Service Invocation

This module contains registered and unregistered web services invoking. One middle-ware, named Web Service Caller Client, has been developed and deployed to discover, select, and invoke either registered or unregistered web services asynchronously. The request and response messages between this client and web service are exchanged in Simple Object Access Protocol (SOAP) format. The output will be processed in this client and the appropriate results will be returned to the browser client.

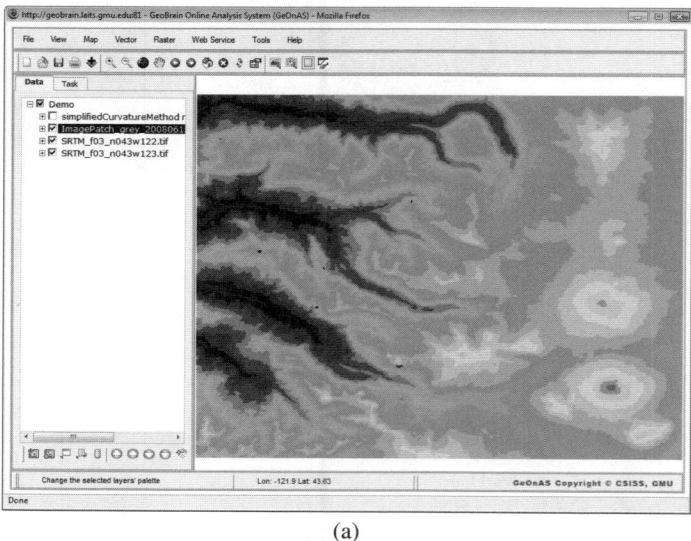

(a)

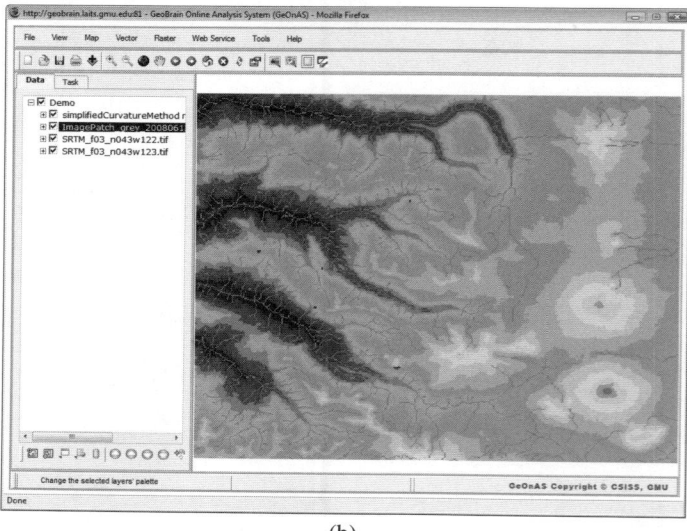

(b)

Fig. 7. Stream Extraction Based on SRTM

Web services registered in CSISS CSW server are organized in the hierarchical structure derived from the semantic types of computation which are defined in ISO 19119:2005, as shown in Figure 6. User could input keywords or click the tree items to search the relevant services. The graphic user interface offers the operations, parameters, data types, binding information, and help for the specified service.

Figure 7 demonstrates the output of the registered stream extraction service which is performed on SRTM dataset.

Because of web services' significant overheads with respects to invocation, E-Mail notification is adopted to inform user the status of web service invocation if he close and save the current project.

4.6 Auxiliary Functions

This module provides Background Scheme, Add to Favorites, Options, User Guide (in pdf and html format), Feedback, FAQ, About Dialog, etc. auxiliary functions.

5 Conclusions and Future Work

According to the statistics from Google Analytics, nearly 3,000 users from USA, Canada, China, Germany, Italy, UK, and other 21 countries and districts have visited GeOnAS since it was released to the education partners for public testing on October 2007. As seen in Fig. 8, the report gives a specific overview map of the site visitors distribution in USA from January 1, 2008 to June 13, 2008. Moreover, GeOnAS was demonstrated its cutting-edge capabilities in geospatial data publishing and accessing, information processing and retrieving, and knowledge building and sharing at AccessData 2008 Workshop[2].

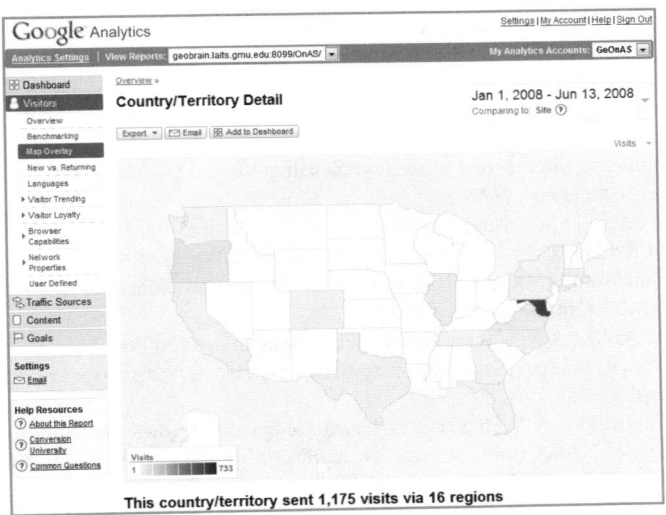

Fig. 8. Visitors Distribution Map

---

[2] http://serc.carleton.edu/usingdata/accessdata/workshop08/team/geobrain.html

Based on the suggestions and comments from users and project partners, we will enrich system functions, tune system performance to be faster and more responsive, and support the Opera and Safari browser in the near future. And the chained web-based services will be implemented to process geospatial datasets in user-defined ways and perform more complex analysis tasks at the end of this year.

In conclusion, GeOnAS not only enhances discovery, retrieval and integration of heterogeneous geospatial data and information, but it also employs distributed web geospatial services to solve a real world task in decision making. We consider that GeOnAS has great potentials to make impacts and differences to the geosciences researchers around the world; and it will help migrate these scientists from desktop GIS or RS software to on-line service-data offers.

Acknowledgements

We acknowledge the support of NASA's Advanced Information System Technology (AIST) program (No. NAG5-13409) and NASA's Research, Education and Applications Solution Network (REASoN) program (No. NNG04GE61A). For providing support for web stream extraction services, we thank Prof. Wei Luo at Northern Illinois University.

References

1. Di, L., Zhao, P., Han, W., Wei, Y., Li, X.: GeoBrain Web Service-based Online Analysis System (GeOnAS). In: Proceedings of NASA Earth Science Technology Conference 2007. 7 pages, CD-ROM. College Park, Maryland (2007)
2. Interactive Visualizer and Image Classifier for Satellites,
   http://www.nsstc.uah.edu/ivics/
3. Zhao, P., Deng, D., Di, L.: Geospatial Web Service Client. In: Proceedings of ASPRS 2005 Annual conference. Baltimore, Maryland, 5 pages, CD-ROM (2005)
4. Global Land Cover Facility, http://glcf.umiacs.umd.edu/index.shtml
5. Tucker, C.J., Grant, D.M., Dykstra, J.D.: NASA's Global Orthorectified Landsat Data Set. Photogrammetric Engineering & Remote Sensing 70(3), 313–322 (2004)
6. Earth Science Gateway, NASA,
   http://esg.gsfc.nasa.gov/web/guest/home
7. Birk, R., Frederick, M., Dewayne, L.C., Lapenta, M.W.: NASA's Applied Sciences Program: Transforming Research Results into Operational Success. Earth Imaging Journal 3(3), 18–23 (2006)
8. Garrett, J.: AJAX: A New Approach to Web Applications (2005),
   http://www.adaptivepath.com/ublications/essays/archives/000385.php
9. Di, L.: GeoBrain - A Web Services based Geospatial Knowledge Building System. In: Proceedings of NASA Earth Science Technology Conference 2004. Palo Alto, California, 8 pages, CD-ROM (2004)
10. Geographic Resources Analysis Support System, http://grass.itc.it

Semantic Analysis for the Geospatial
Web – Application to OWL-DL Ontologies

Alina-Dia Miron, Jérôme Gensel, and Marlène Villanova-Oliver

Laboratoire d'Informatique de Grenoble, 681 rue de la Passerelle BP 72,
38402 Saint Martin d'Hères Cedex, France
{Alina-Dia.Miron,Jerome.Gensel,Marlene.Villanova-Oliver}@imag.fr

Abstract. Towards the concretization of Egenhofer's idea of a future Geospatial Semantic Web, an important limitation is the absence of spatial and temporal reasoners capable of combining quantitative and qualitative knowledge for inferring new relations between individuals described in OWL or RDF(S) ontologies. As a first response to this limitation, we propose the use of the ONTOAST system, a spatio-temporal ontology modeling and semantic query environment compatible with OWL-DL. In this paper we illustrate the practical use of ONTOAST for querying the Geospatial Semantic Web in order to discover *semantic associations* between individuals. We present problems we have encounter when handling spatial and temporal knowledge described in OWL-DL as well as the solutions we have adopted.

Keywords: Geospatial Semantic Web, ontology, semantic analysis, spatiotemporal reasoning, context definition.

1   Introduction

Nowadays, one of the most popular Web usages is information search. Current query systems and search engines retrieve relevant Web documents by applying syntactic matching between given keywords and textual content of Web documents. But with the tremendous growth of the amount of digital data available on the Web, problems of data relevance and information overload become acute. A second generation Semantic Web [1] addresses those issues and promises to increase the performances and the relevance of search engines, by annotating the content of any Web resource with machine understandable ontological terms. While many of the requirements for capturing semantics and expressing ontologies are successfully addresses by the Semantic Web initiative, there is still a fundamental lack when considering the existing standardized descriptions [2] and reasoning mechanisms [3][4] dealing with spatial and temporal information. The nature of geospatial reasoning being highly mathematical, the logical formalisms behind the ontology languages recommended by W3C (RDF(S), OWL) prove to be unsuited for handling such data. As a consequence, the processing of spatial and temporal information needs to be performed using external formalisms and tools, but the results of such computations must be integrated into the ontological space, in order to be made available to different reasoners for query answering.

M. Bertolotto, C. Ray, and X. Li (Eds.): W2GIS 2008, LNCS 5373, pp. 37–49, 2008.

We propose the use of the ONTOAST system [5], as an answer to the absence of specialized spatial and temporal inference engines defined on top of OWL or RDF(S) ontologies. ONTOAST is a spatio-temporal ontology modeling and semantic query environment, able to reason about spatial, temporal and thematic knowledge and which can be used in the Semantic Web context through its compatibility with OWL-DL [6]. The spatio-temporal reasoning capabilities of ONTOAST rely on both a powerful and extensible typing system and an Algebraic Modeling Language [7] which offers support for complex algebraic computations and facilitates numerical data transformation into qualitative relations.

The ONTOAST representation model is flexible enough to handle the coexistence of quantitative spatial and temporal data, in the form of exact geometries and time instants or intervals, and imprecise data in the form of qualitative spatial and temporal relations. The inferred qualitative relations, obtained from existing numerical and/or qualitative information, as showed in [5], can be easily integrated within an OWL ontology and made available to classifiers for further reasoning.

We illustrate in this paper the use of ONTOAST in the field of *semantic association* discovery [8][9][10]. This relatively new research area focuses on answering semantic queries like *"Is instance x in any way connected to instance y?"*, by retrieving all paths which connect *individuals[1]* x and y within the considered ontology graphs. In this paper, we propose a *Semantic Association Discovery Framework* that uses the spatial and temporal reasoning capabilities of ONTOAST for restraining the semantic search space and for inferring new *semantic associations*. In the first case, the ONTOAST reasoner filters the ontological knowledge with respect to a spatial and a temporal query contexts specified by the user. In the second case, based on spatial information describing OWL *individuals* and *object properties*, ONTOAST can infer implicit and/or new qualitative spatial relations. Those relations are then used for constructing new and possibly interesting semantic associations, as showed in section 5.4. We also describe in this paper the meta-ontology that we use for bridging the geometric descriptions usually attached to spatial resources on the Web, and the definitions of qualitative spatial relations linking them. Temporal information is managed through the use of special stamps that model the temporal validity of object properties in OWL and which can be handled by the ONTOAST reasoner.

The paper is organized as follows: in section 2 the semantic analysis field is presented and in section 3 the adaptation of the ρ-path definition to the OWL-DL representation structures is discussed. Section 4 illustrates the semantic framework we propose for handling spatial and temporal knowledge for semantic association discovery. The query approach is illustrated in Section 5 as well as the related algorithms. Section 6 deals with related work and Section 7 concludes.

2 Context

So far, research in the field of Semantic Web has mainly focused on areas such as information integration and interoperability, semantic search and browsing... However,

---

[1] The terms *individual* and *object* are used in this paper as synonyms, for designating an instance of a concept.

while these issues are obviously important, the discovery of semantics itself can not be ignored since the analysis process it requires, although complex, makes possible the transition from data towards knowledge. Discovery of semantics refers to a semi-automatic reasoning process which infers implicit and/or new semantic relations from existing ontological knowledge. In this direction, the *semantic analysis* [8] has recently emerged as a new research field, which aims at the automatic identification of *semantic associations* linking entities together, within semantically annotated data. The concept was first introduced by [8] and formally defined [9][10] as a sequence x, p_1, e_1, p_2, ... e_{n-1}, p_n, y of RDF entities (x, e_1, ..., e_{n-1}, y) linked by RDF properties (p_1, ...p_n). *Semantic association* discovery has already been successfully used in domains such as homeland security, biomedical patents retrieval, or conflict of interest detection...

However, the choice of using RDF(S) as an ontology modeling language can be considered as too restrictive given its limited expressive power when compared to more powerful Description Logics or Object Knowledge Representation Languages. This observation has motivated our previous work [11] whose goal was to adapt the semantic analysis techniques to OWL-DL ontologies. Our idea was, on the one hand, to obtain more expressive power for defining complex ontologies and, on the other hand, to perform extended inferences that explore, for example, OWL-DL axioms or ontology alignments. This work has been done without considering any domain of application in particular.

In this paper, we propose to emphasis both the spatial and the temporal dimensions of information often contained by OWL ontologies. We show how spatial and temporal information described in Semantic Web documents can also be exploited by the ρ-*path* discovery algorithms. As a matter of fact, *semantic analysis* has mainly focused on thematic metadata, which holds information about how entities are related in different domains. Still, given the increasing popularity of spatial Web application and following Egenhofer's idea of a future Geospatial Semantic Web [3], we claim that taking into account both the spatial and temporal dimensions of data during the semantic analysis process becomes necessary. A spatial, temporal and thematic ρ-*path* discovery framework would support the inference of new and possibly relevant *semantic associations* (ρ-*path*) between individuals. For instance, based on an ontological description about two persons who live in the same neighborhood for *ten* years, the spatio-temporal ρ-*path* discovery framework may deduce that the two persons are very likely to know each other. So, the inferred proximity of their houses can be automatically taken into account for suggesting that there might be a link between those two persons.

3   Semantic Associations in OWL-DL

The *semantic association* discovery process has been defined for graph data models [9]. Graphs are a popular and simple knowledge representation paradigm, which offer a natural and intuitive way to model *relations* between *individuals*. OWL being built on top of the RDF(S) graph model, translating the *semantic analysis* in the context of OWL ontologies is straightforward.

We consider an OWL vocabulary V, formally defined [14] as consisting of a set of literals V_L and seven sets of URI references: V_C, V_D, V_I, V_{DP}, V_{IP}, V_{AP}, and V_O, where

V_C is the set of *class* names and V_D is the set of *datatype* names of a vocabulary. V_{AP} represents the *annotation property* names, V_{IP}, the *individual-valued property* names, V_{DP}, the *data-valued property* names, V_I, the *individual* names and V_O, the *ontology* names of a vocabulary. The same source [14] defines an OWL interpretation as a tuple of the form: $I=<R, EC, ER, L, S, LV>$, where R is the set of resources of I, LV represents the literal values of R and L is an interpretation that provides meaning to typed literals. The mapping S provides meaning for URI references that are used to denote OWL individuals, and helps provide meaning for annotations. We also consider O ($O \subset R$, $O \cap LV = \varnothing$) as being a non empty set of *class instances (objects)*, the set of *object property instances* Γ ($\Gamma \subseteq 2^{O \times O}$) and the set Λ containing the *datatype property instances* ($\Lambda \subseteq 2^{O \times LV}$). Given the fact that we search for links between individuals in an OWL concrete model we are only interested by a subset of the mappings EC and ER defined for the interpretation model I. Thus, we define the function Ext_C ($Ext_C: V_C \rightarrow 2^O$) as being a specialization of the EC mapping for providing meaning for URI references that are used as OWL *classes* ($Ext_C(Cl) \subseteq 2^O$, $Cl \in V_C$). We also consider $Ext_R: V_{IP} \rightarrow \Gamma$, ($Ext_R(Rel) \subseteq \Gamma \subseteq 2^{O \times O}$, $Rel \in V_{IP}$) and $Ext_D: V_{DP} \rightarrow \Lambda$, ($Ext_D(Att) \subseteq \Lambda$, $Att \in V_{DP}$), as two sub mappings of ER which provide meaning for URI references that are used as OWL *object properties* respectively OWL *datatype properties*. In the reminder of this paper we denote an *object property* instances as a triple of the form (Rel, o_1, o_2), which states that the relation Rel holds between individuals o_1 and o_2, where $Rel \in V_{IP}$, $o_1, o_2 \in V_O$, $S(o_1) \in O$, $S(o_2) \in O$, $(S(o_1), S(o_2)) \in Ext_R(Rel)$. We also refer to object properties instances by the name of *tuples*.

$$G=(V_G, E_G) \; where \; V_G = V_C \cup V_D \cup V_I \cup V_L \cup O \cup LV, E_G = V_{IP} \cup V_{DP} \cup \Gamma \cup \Lambda,$$
$$E_G \subseteq \{(u,v) | u \in V_C, v \in V_C \cup V_D, \exists Rel \in V_{IP} : u \in Domain(Rel) \wedge v \in Range(Rel)\} \cup$$
$$\left\{ (i,j) \middle| i \in O, j \in O \cup LV, \exists Rel \in V_{IP} : (S(i), S(j)) \in Ext_R(Rel) \vee \atop \exists Att \in V_{DP} : (S(i), S(j)) \in Ext_D(Att) \right\}. \tag{1}$$

For each OWL ontology Ω, we can build a directed graph G defined by equation (1), where V_G is a set of vertices, and E_G is a set of oriented edges between the vertices. In this context, we say that two individuals x and y ($x, y \in O$) are *semantically associated* if, in the considered ontology graph G, there exists at least one sequence, alternating *individuals* linked by different *tuples*, that connects x to y. In other words, one states the existence of at least one graph *path* that begins with node x and passes by a set of *intermediary individuals* ($o_i \in O, i \in 1..n$) linked by *tuples* ($(R_i, o_{i-1}, o_i) \in \Gamma$, $R_i \in V_{IP}$, $i \in 1..n$) and ends by the node y (see equation (2)). For describing a ρ-path in a unique way, in this paper, we use the following notation:

$$\rho - path(x,y) = x \overset{R_1}{\rightarrow} o_1 \overset{R_2}{\rightarrow} ... \overset{R_{n-1}}{\rightarrow} o_{n-1} \overset{R_n}{\rightarrow} y, \; o_i \in O, 0 \leq i \leq n, \; x = o_0, y = o_n,$$
$$R_j \in V_{IP}, (R_j, o_{j-1}, o_j) \in \Gamma, 1 \leq j \leq n. \tag{2}$$

Concretely, in the simple ontology illustrated by Fig.1, three ρ-paths exist between *pers1* and *pers3*: ρ-path p is direct since it connects *pers$_1$* to *pers$_2$* using a single *tuple* of the relation *studentOf*, q and r are indirect since they pass through one respectively two intermediary objects. Their structures are detailed bellow:

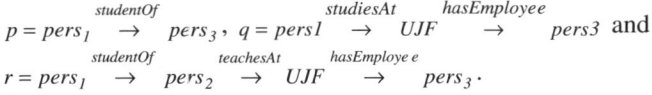

$$p = pers_1 \xrightarrow{studentOf} pers_3, \; q = pers1 \xrightarrow{studiesAt} UJF \xrightarrow{hasEmployee} pers3 \text{ and}$$
$$r = pers_1 \xrightarrow{studentOf} pers_2 \xrightarrow{teachesAt} UJF \xrightarrow{hasEmploye e} pers_3.$$

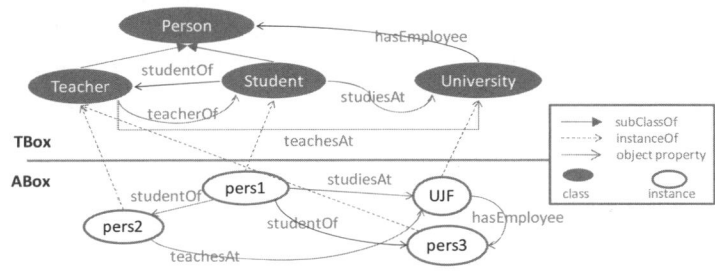

Fig. 1. Example of simple OWL ontological description

In the ontological graphs as previously presented, we are interested in mining techniques that are able to deduce implicit spatial and temporal relations between individuals which can lead to the discovery of new *ρ-paths*. The semantic analysis framework we propose is illustrated in section 4.

4 Geospatial and Temporal Semantic Analysis Framework

Our proposal is based on the use of ONTOAST (which stands for ONTOlogies in Arom-ST) as a spatio-temporal ontology modeling and semantic query environment. ONTOAST [6] is an OWL-DL compatible extension of the Object Based Representation System AROM [12], a generic tool designed for knowledge modeling and inference. The originality of this system comes from its powerful and extensible typing system and from its Algebraic Modeling Language [7] used for expressing operational knowledge in a declarative way. This language allows one to specify the value of a *variable* using numerical and symbolic equations between the various elements composing the knowledge base. ONTOAST is built upon the spatio-temporal module of AROM. The interest of using ONTOAST lies in its predefined set of qualitative spatial and temporal *associations* (presented in details in [5]). Those *associations* can be used in order to complete data on the modeled *objects* as well as to allow a more flexible query formulation. Thus, ONTOAST ontologies are flexible enough to handle the coexistence of, on the one hand, quantitative spatial and temporal data in the form of exact geometries and time intervals or instants and, on the other hand, imprecise data in the form of qualitative spatial and temporal relations. These two kinds of information complement one another and offer advanced reasoning capabilities. ONTOAST takes into account three categories of qualitative spatial relations: *topology*, *orientation* and *distance*. They can be automatically inferred from existing knowledge when they are needed, or explicitly defined by users. In order to perform similar inferences on temporal data, ONTOAST manages a set of qualitative temporal relations (*before, after, starts/started-by, finishes/finished-by, during/contains, equals, meets /met-by, overlaps/overlapped-by*).

This paper presents the use of ONTOAST in the *semantic association* discovery process. As illustrated in Fig. 2, the *Semantic Analysis Framework* contains five main modules respectively in charge of the *Knowledge Acquisition*, the *Query Interface*, the *P-path Discovery*, the *Result Classification* and the *Result Visualization*. They are organized as distinct but communicating exploitation tools, on top of the ONTOAST System. In order to use the reasoning facilities provided by ONTOAST, the ontological knowledge has to be translated into the AROM formalism and stored into a local object-oriented *Knowledge Repository*.

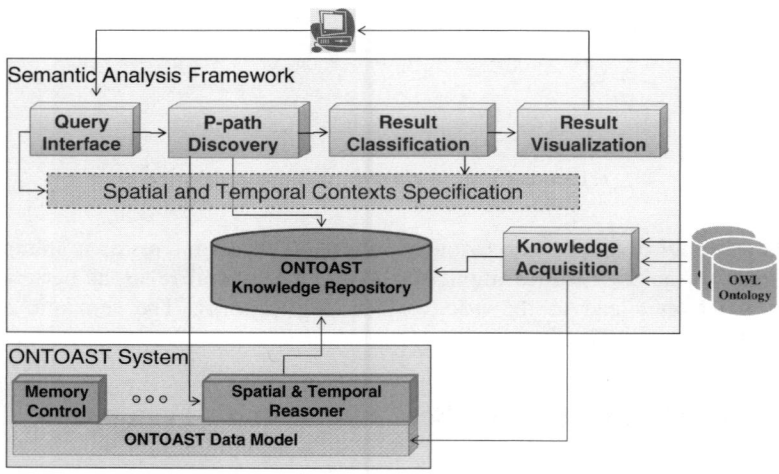

Fig. 2. Overview of ONTOAST Framework for the discovery of *semantic associations*

Once the ontological knowledge is imported into the *ONTOAST Knowledge Repository*, spatial and temporal semantic inferences can be activated. During the *ρ-path* discovery process, the *Knowledge Repository* data are filtered using the *Spatial and Temporal Contexts Specification*, in order to reduce the search space. The filtered ontological knowledge is exploited by the *ρ-path Discovery* module which searches for the semantic associations between two individuals. The *Spatial and Temporal ONTOAST Reasoner* is used both in the filtering phase (for inferring spatial characteristics and temporal validity for individuals) and in the *ρ-path Discovery* phase (for inferring[2] spatial connections between individuals). The obtained *ρ-paths* are then classified by the *Result Classification* module using the context specifications. It is clear that the handling of spatial and temporal information increases the scope of the *semantic analysis*, but raises new representation and reasoning challenges. For instance, the limited typing system adopted by RDF(S) and OWL does not offer support for spatial extensions. As a consequence, in order to model spatial data in OWL, dedicated concepts which simulate spatial datatypes have to be used. For example, a *polygon*

---

[2] We refer to *ρ-path* construction when qualitative spatial relations are inferred and added to the knowledge base, allowing the discovery of new *ρ-paths*.

(instance of a *Polygon* class) will be represented by a list of *Points* (*x,y*) objects connected to a given *Coordinate System.*

Several geospatial ontologies which model geometric features have been proposed so far. An assessment study which compares 45 geospatial and temporal ontologies relevant to geospatial intelligence, from the annotation, qualitative reasoning and information integration points of view, has recently been published [13]. Following recommendations of the authors we have chosen the *GeoRSS-Simple*[3] OWL encoding as a reference spatial ontology. *GeoRSS-Simple* is a very lightweight format for spatial data that developers and users can quickly and easily integrate into their existing applications or that they can use for expressing semantic annotations with little effort. It supports basic geometries (*point, line, box, polygon...*) and covers the typical use cases when encoding locations. Then, the ONTOAST *Knowledge Acquisition Module* is constructed to recognize spatial data types modeled using ontological concepts in the *GeoRSS-Simple* ontology.

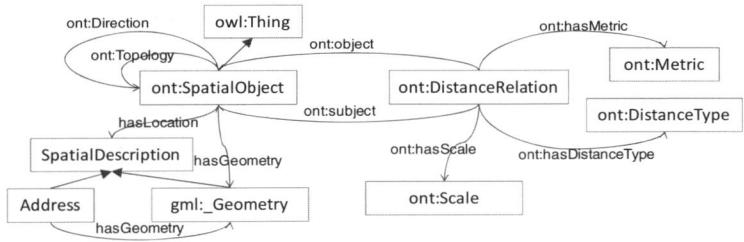

Fig. 3. General OWL ontology defining Spatial Objects. It integrates the *QualitativeSpatialRelations* ontology (*ont* prefix) we have chosen as a reference ontology for modeling qualitative spatial relations compatible with ONTOAST, and the *GeoRSS* ontology (*gml* prefix) our reference for modeling geometric features.

In order to represent in OWL qualitative spatial relations that can be handled by the ONTOAST reasoner, we propose the use of *QualitativeSpatialRelations*[4] ontology. It contains two generic types of spatial relations, *Direction* and *Topology,* modeled using *object properties*. The *Direction* relation has nine specializations which allow the expression of cardinality directions (*Nof, Sof, Eof, Wof, CenterOf, NEof, NWof, SEof, SWof*) existing between spatial objects. Specializations of the *Topology* relation have also been defined: *Disjoint, Contains, Crosses, Touches, Equals, Intersects, Overlaps* and *Within.* Distance relations are harder to represent since they come with attributes specifying, for example, the metric system employed when calculating the distance (Euclidian distance, shortest road, drive time, etc.), the scale, or the distance type when regions of space are described (average border points, gravity centers, administrative centers, etc.). Since OWL-DL does not allow object properties to have some attributes, distances between spatial objects are reified as objects of the *DistanceRelation* (see Fig. 3). For specializations (*VeryFar, Far, Close,* and *VeryClose*)

---

[3] The ontology can be found at: http://mapbureau.com/neogeo/neogeo.owl
[4] The ontology can be found at: www-lsr.imag.fr/users/Alina-Dia.Miron/
QualitativeSpatialRelations.owl

of the *DistanceRelation* allow the specification of absolute distances between concepts. Fig. 3 illustrates the general OWL ontology we have used for annotating spatial
objects. We have considered that spatial information can be attached to *Spatia-
lObjects* in three ways: a) using the *hasGeometry* property which designates a concrete *Geometry* for the specified geographic object, b) using the *hasLocation* property
that can refer to an *Address* or a concrete *Geometry*, c) through the use of a qualitative
spatial relation.

In order to express temporal data in OWL, one can exploit several dedicated
datatypes: *xsd:dateTime*, *xsd:date*, *xsd:gYearMonth*, *xsd:gMonthDay*, *xsd:gDay*,
xsd:gMonth. These datatypes will be used for associating temporal characteristics
with individuals but cannot be used for expressing *tuples* validity within the decidable
frontiers of OWL-DL. In OWL-Full, object properties can be used as classes and thus,
their instances can be described by data properties. Considering that ONTOAST is not
compatible with OWL-Full, we cannot adopt this solution. An interesting approach is
presented in [15] and consists in stamping RDF triples with instants or intervals of
time. Since current OWL recommendations do not support temporal stamps, we propose to simulate them through special annotations managed by the ONTOAST
Knowledge Acquisition Module. Three examples of temporal annotations are shown
in Fig.4. The first two assign temporal intervals to *tuples* (instances of object properties) and the last one stamps the correspondent *tuple* with a temporal instant.

```
ObjectPropertyAssertion(Comment("temporalValidity(2005-10-10,T10:30:00 -)") knows7 pers1 pers2)
ObjectPropertyAssertion(Comment("temporalValidity(2003-01-10,2006-30-06)") studiedAt  pers1 UJF)
ObjectPropertyAssertion(Comment("temporalValidity(2008-08-10)") colaboratesWith pers1 pers2)
```

Fig. 4. Examples of temporal annotations on OWL DL *tuples*[5]

5   Querying Approach

Finding *semantic associations* between two individuals can be achieved by applying a
recursive algorithm of graph path discovery (see section 5.1) between two given vertices. In the open Geospatial Semantic Web environment, those algorithms can return
a high number of results. It is therefore necessary to establish some filter criteria that
limit the search space according to the user's interests. In the following sections two
such filters are presented. To illustrate our approach for the discovery of geospatial
and temporal semantic association, in the reminder of this section we refer to an ontology describing relations of collaboration existing between researchers (Fig.5). Let
us consider a query that searches for the existing links between $pers_2$ and $pers_4$.

5.1   *P-path* Discovery Algorithm

The *P-path Discovery* module of the *Semantic Analysis Framework* presented in section 3 integrates the depth first discovery algorithm presented in Fig.6. The *ρ-paths*

---

[5] The syntax used for this example is the Functional-Style Syntax defined for OWL 2.

between two given individuals x and y are inferred using an advancement stack (called *stack*) and a result stack, *pPath,* both containing intermediate *tuples.* They are initialized with a virtual tuple, having as objects the start individual (see lines 2-3). At each step of the algorithm, if advancement in the graph if possible, all the *tuples* having as subject the current individual (*source*) and which satisfy the context specifications are added to the *stack* (see lines 5-9). If, on the contrary, there is no *tuple* having as subject the current *source*, the last considered *tuple* is eliminated from the *pPath* and from the work stack as well. When adding an intermediate *tuple* to the *pPath,* the absence of cycles and the length constraint (L_{Max}) are checked (line 15). If the current *tuple* t, blocks the construction of a valid *p-path*, t is eliminated (lines 24-28) both from the *stack* and from the *pPath.* The algorithm is executed as long as the stack contains possible path alternatives (line 29). In order to add a *tuple* t to the current *p-path* its elements must satisfy the *Temporal* and *Spatial Context* specifications. Those tests are performed by the *inCtx* function (see line 9). The set of inferred *p-paths* is kept by the variable *resultSet* which is a list of lists (see lines 18, 30).

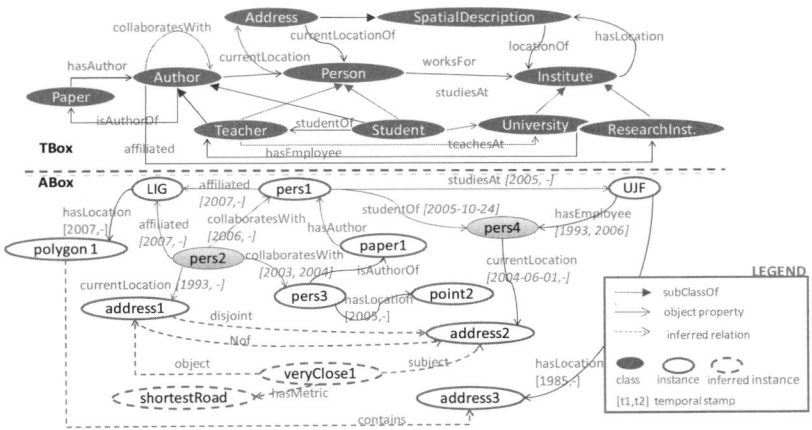

Fig. 5. The OWL ontology modeling a small social network in the research domain

In the worst case scenario, when all objects in G are connected to each other by a maximum of d_{max} direct *tuples,* the maximum complexity of this algorithm is $O(d_{max}^{Lmax+1})$. This case is highly improbable since the instances of object properties described in ontologies rarely hold between all pairs of objects. Moreover, the use of contexts considerably limits the computations. Nevertheless, we intend to optimize this first version of algorithm, following the approach presented in [16], which provides promising fast algorithms for solving path problems. In the case when no temporal and spatial filters are applied, the algorithm exhaustively discovers all paths linking $pers_2$ to $pers_4$ in the ontology of Fig.5. Results obtained in this case are illustrated in Table 1.

```
pPathDiscovery(start, stop, Lmax)                              and not cycle(pPath,getObject(t))
  1.   source=start                                    16.        push(pPath, t)
  2.   push(pPath, create Tuple (null, null, source))  17.        if getObject(t)=stop
  3.   push(stack, create Tuple (null, null, source))  18.            add(resultSet, pPath)
  4.   do                                              19.            pop(pPath)
  5.     if hasTuples(source)                          20.            pop(stack)
  6.       if getSubject(top(stack))=source            21.            source=getSubject(top(stack))
  7.         for each tpl in tuples(source)            22.        else source=getObject(t)
  8.           if inCtx(tpl)                           23.      else
  9.             push(stack, tpl)                       24.        if contains(pPath,t)
 10.     else                                          25.          pop(pPath)
 11.       pop(pPath)                                  26.        if size(stack) >1
 12.       pop(stack)                                  27.          pop(stack)
 13.       source=getObject(top(stack))                28.          source= getSubject(top(stack))
 14.     t=top(stack)                                  29.   while size(stack)>1
 15.     if size(pPath)≤Lmax and not contains(pPath,t) 30.   return resultSet
```

Fig. 6. The extended *ρ-path* discovery algorithm

Table 1. *P-paths* linking *pers₂* and *pers₄* in the ontology illustrated by Fig.6

$$p_1 = pers2 \xrightarrow{collaboratesWith} pers1 \xrightarrow{studentOf} pers4$$

$$p_2 = pers_2 \xrightarrow{collaboratesWith} pers_1 \xrightarrow{studiesAt} UJF \xrightarrow{hasEmployee} pers_4$$

$$p_3 = pers_2 \xrightarrow{collaboratesWith} pers_3 \xrightarrow{isAuthorOf} paper_1 \xrightarrow{hasAuthor} pers_1 \xrightarrow{studentOf} pers4$$

$$p_4 = pers_2 \xrightarrow{collaboratesWith} pers_3 \xrightarrow{isAuthorOf} paper_1 \xrightarrow{hasAuthor} pers_1 \xrightarrow{studiesAt} UJF \xrightarrow{hasEmployee} pers_4$$

5.2 The Temporal Context

The first type of context we consider for a query is the *temporal context C_{temp}*. It acts as a filter and narrows down the search area to *ρ-paths* that meet a certain qualitative relation with a given temporal interval or instant. The *temporal context* is specified using an interval of the type *(start_date, end_date)* or *(date, duration)*, together with a topological relation. The implicit topological relation is the *inclusion*, but all the topological temporal AML operators defined in [5] can be used: *before/after, starts/started-by, finishes/finished-by, during/contains, equals, meets/met-by, overlaps/overlapped-by*. If a time interval is specified through a pair *(date, duration)*, the temporal attributes describing *tuples* or *objects* are compared to the interval *[date-duration, date+duration]*. The implicit temporal interval starts with the beginning of time and holds up to the present moment, but it can be adapted to the user's preferences.

When analyzing an instance *e* (*tuple* or *object*) candidate for its integration into a *ρ-path*, the system compares its validity, given by its temporal data, specified by *data properties* (for *individuals*) or *annotation stamps* (for *tuples*), to the *temporal context C_{temp}*. If the entity has no temporal attributes, the system considers that it has always existed and that it is eternal. For the example shown in Fig. 5, considering the temporal context C_{temp}=([2005, 2008], *overlaps*), will eliminate from the resulting *ρ-paths* p_3 and p_4 (see Fig. 5). The reason is the validity of the *tuple* (*collaboratesWith pers₂ pers₃*) given by its temporal stamp: [2003, 2004]. The interval [2003, 2004] does not satisfy the *overlaps* relation with the C_{temp} interval [2005, 2008], so the *ρ-paths* construction cannot be completed.

5.3 The Spatial Context

The ONTOAST spatial reasoner can be used to filter and eliminate from the set of results the *ρ-paths* which contain *intermediate objects* or/and *tuples* that do not satisfy a given qualitative spatial relation with a specified region of the geographical space. We define the spatial query context $C_{spatial}$, as the pair *(spatial object, qualitativeSpatialRelation)*, where *qualitativeSpatialRelation* is one of the spatial relations described in section 4. Similarly to the case of the temporal context, during the analysis of an *instance e* candidate for integration into a *ρ-path p*, the system compares its spatial characteristics given by its geometry or inferred by the system (in the case of *objects*), to the spatial context $C_{spatial}$. If *e* is not consistent with $C_{spatial}$, it will not be taken into account as a possible intermediary object in a *ρ-path*. The *instances* which have no associated spatial extent are considered as being consistent with the spatial context $C_{spatial}$.

Let us consider that, in the example illustrated in Fig.5, *address₁* and *address₂* describe two places in Paris at one street away from each other, *address₃* is located in Versailles, *polygon₁* describes the geometric limits of all the sites administrated by the Computer Science Laboratory from Paris and *point₂* represents geospatial coordinates in England. Now, if the user is only interested in connections between researchers from the French region Ile de France he/she imposes as a spatial filter for the semantic query $C_{spatial}$=*(IleDeFrance, contains)*. *IleDeFrance* must be a spatial object defined within the accessible ontologies. *Pers₃,* whose current location (*England*) does not satisfy the inclusion relation becomes inconsistent with the specified context. In consequence, the construction of *ρ-paths* passing by *pers₃* (*p₃* and *p₄* from Table 1) cannot be completed.

5.4 Spatial Inference for *P-Path*

The ONTOAST spatial reasoner can also be used by an extended *ρ-path* discovery algorithm which aims at identifying individuals who can be linked through a relevant spatial relation that stands between them. The algorithm has the following steps:

1. If *x* is not a spatial object or if its geometry is unknown, the system builds the set S_x containing the spatial objects that are related to *x* by *ρ-paths* $α_1, α_2, ..., α_z$ of a given maximum length l_{Max}: $|α_i| \leq l_{Max}$ $\forall i \in 1..z$ and which are consistent with the specified contexts. For a spatial object *x* with known geometry the set S_x exclusively contains the object itself.
2. The same steps are taken for building the set S_y.
3. Among the objects contained by the sets $S_x \backslash S_y$ and S_y, the system exhaustively infers the existence of qualitative spatial relations. The qualitative relations inferred this way are added up to the ontology and the newly obtained *ρ-paths*: $α_i \rightarrow spatial\_relation \leftarrow β_j$ will be taken into account as a result of the semantic analysis.

Using this algorithm, new ρ-paths can be discovered. For example, in the ontology of Fig. 5, after building the sets $S_{pers2}=\{address_1,\ polygon_1\}$ and $S_{pers4}=\{address_2,\ address_3\}$ the ONTOAST spatial reasoner checks which qualitative spatial relations exist between pairs: $address_1$ and $address_2$, $address_1$ and $address_3$, $polygon_1$ and $address_2$, $polygon_1$ and $address_3$. Let us focus on the possible inferences from the first pair. Several geocoding Web Services exist to this day (Yahoo!Maps, Google Maps, MapPoint, ...), that support address transformation into corresponding latitude and longitude coordinates. So, obtaining geographic positions from address specifications is relatively easy. The obtained quantitative spatial information will be used by the ONTOAST spatial reasoner for inferring, through geometric computations, the spatial relations existing between addresses. Knowing that $address_1$ is in a close proximity of and at Nord from $address_2$, the tuples $veryClose(address_1,\ address_2)$, $Nof(address_1,\ address_2)$ and $disjoint(address_1,\ address_2)$ will be added up to the ontology, which facilitates then the discovery of three new ρ-paths:

$$p_5 = pers_2 \xrightarrow{currentLocation} address_1 \xrightarrow{object} veryClose_1 \xrightarrow{subject} address_2 \xrightarrow{currentLocation} pers_4;$$

$$p_6 = pers_2 \xrightarrow{currentLocation} address_1 \xrightarrow{Nof} address_2 \xrightarrow{currentLocation} pers_4;$$

$$p_7 = pers_2 \xrightarrow{currentLocation} address_1 \xrightarrow{disjoint} address_2 \xrightarrow{currentLocation} pers_4;$$

6 Related Work

Besides the related work previously cited, another interesting approach is presented in [17]. The authors propose the integration of RDF ontologies and ORACLE databases, through the use of ORACLE Semantic Data Store. The latter provides the ability to store, infer and query semantic data in the form of simple RDF descriptions or RDFS ontologies. Spatial attributes attached to ontological objects, modeled in RDF using coordinates lists (for example, instances of the *Point* class), are translated into values of a spatial datatype in ORACLE (SDO_GEOMETRY). Those translations are very useful as all the ORACLE Spatial query facilities, including its topological and proximity operators, are available. However, this approach does not make it possible to combine qualitative and quantitative reasoning. In addition, in order to query a database, it is necessary to know in advance the relationships that hold between the *individuals*. Translated into the *semantic analysis* vocabulary, it means that the composition of the ρ-paths must be known *a priori* and spatial relations will only be used for filtering the results.

7 Conclusions and Perspectives

In this paper, we have studied ways of improving the analytical power of *semantic associations*. We proposed a means for exploiting the temporal and spatial dimension of Semantic Web resource descriptions in order to discover new ρ-paths linking two given individuals. Using ONTOAST as a spatial reasoner for OWL ontologies, new inferences can be produced that lead to new and possibly helpful insights into the ways in which individuals are connected within ontological domains.

We currently consider the possibility of introducing a new module of *Trust management* into the *Semantic Analysis Framework*. Its purpose would be, on the one hand, to filter the ontological knowledge and, on the other hand, to communicate with the *Result Classification Module* for determining the weights of the discovered semantic associations.

As a primary future goal, we intend to implement a prototype of the *Semantic Analysis Framework* and to test our algorithms on an extended ontological base in order to quantify their relevance and performance in real world semantic association discovery scenarios. We also plan a detailed performance study using large synthetic OWL datasets.

References

1. Berners-Lee, T., Hendler, J.A., Lassila, O.: The Semantic Web. Scientific American 284(5), 34–43 (2001)
2. Agarwal, P.: Ontological Considerations in GIScience. International Journal of Geographical Information Science. 19(5) (2005)
3. Egenhofer, M.J.: Towards the Semantic Geospatial Web. In: GIS 2002. Virginia, USA (2002)
4. O'Dea, D., Geoghegan, S., Ekins, C.: Dealing with Geospatial Information in the Semantic Web. In: Australian Ontology Workshop. Sydney, Australia (2005)
5. Miron, A., Gensel, J., Villanova-Oliver, M., Martin, H.: Towards the geo-spatial querying of the semantic web with ONTOAST. In: Ware, J.M., Taylor, G.E. (eds.) W2GIS 2007. LNCS, vol. 4857, pp. 121–136. Springer, Heidelberg (2007)
6. Miron, A., Capponi, C., Gensel, J., Villanova-Oliver, M., Ziébelin, D., Genoud, P.: Rapprocher AROM de OWL..., Langages et Modèles à Objets, in French, Toulouse, France (2007)
7. Moisuc, B., Davoine, P.-A., Gensel, J., Martin, H.: Design of Spatio-Temporal Information Systems for Natural Risk Management with an Object-Based Knowledge Representation Approach. Geomatica 59(4) (2005)
8. Sheth, A., Arpinar, B.I., Kashyap, V.: Relationships at the Heart of Semantic Web: Modeling, Discovering, and Exploiting Complex Semantic Relations. In: Enhancing the Power of The Internet: Studies in Fuzziness and Soft Computing, Springer, Heidelberg (2002)
9. Anyanwu, K., Sheth, A.: ρ-Queries: Enabling Querying for Semantic Associations on the Semantic Web. In: WWW2003, Budapest, Hungary (2003)
10. Halaschek, C., Aleman-Meza, B., Arpinar, B.I., Sheth, A.P.: Discovering and Ranking Semantic Associations over a Large RDF Metabase. In: Proceedings of the 30th VLBD Conference, Toronto, Canada (2004)
11. Miron, A.D., Gensel, J., Villanova-Oliver, M.: Thematic and Spatio-Temporal Semantic Analysis for OWL-DL. Internal Report, LIG, Grenoble, France (2008)
12. Page, M., Gensel, J., Capponi, C., Bruley, C., Genoud, P., Ziébelin, D., Bardou, D., Dupierris, V.: A New Approach in Object-Based Knowledge Representation: the AROM System. In: Monostori, L., Váncza, J., Ali, M. (eds.) IEA/AIE 2001. LNCS, vol. 2070, pp. 113–118. Springer, Heidelberg (2001)
13. Ressler, J., Dean, M.: Geospatial Ontology Trade Study. In: Ontology for the Intelligence Community (OIC 2007), Columbia, Maryland (2007)
14. Patel-Schneider, P., Hayes, P., Horrocks, I.: OWL Web Ontology Language Semantics and Abstract Syntax. W3C Recommendation (2004)
15. Gutierrez, C., Hurtado, C., Vaisman, A.: Temporal RDF. In: Gómez-Pérez, A., Euzenat, J. (eds.) ESWC 2005. LNCS, vol. 3532, pp. 93–107. Springer, Heidelberg (2005)
16. Tarjan, R.: Fast Algorithms for Solving Path Problems. J.ACM 28(3), 594–614 (1981)
17. Perry, M., Sheth, A.: A Framework to Support Spatial, Temporal and Thematic Analytics over Semantic Web Data. Technical Report KNOESIS-TR-2008-01 (2008)

Automatic Transformation from Semantic Description to Syntactic Specification for Geo-Processing Service Chains

Peng Yue[1], Jianya Gong[1], and Liping Di[2]

[1] State Key Laboratory of Information Engineering in Surveying, Mapping and Remote Sensing, Wuhan University, 129 Luoyu Road, Wuhan, China, 430079
[2] Center for Spatial Information Science and Systems (CSISS), George Mason University Suite 620, 6301 Ivy Lane, Greenbelt, MD 20770
geopyue@gmail.com, geogjy@163.net,ldi@gmu.edu

Abstract. In a service-oriented environment, where large volumes of geospatial data and diverse geo-processing functions are accessible as services, it is necessary to create service chains as geo-processing workflows to solve complex geospatial problems. Approaches developed in the literature for chaining services can be classified into two categories: static and dynamic. Industry-wide standards such as BPEL provide languages for syntactic specification of service chains. They can not support dynamic service chaining directly, thus can be classified as static approaches. Semantic Web technologies have been widely used to enable dynamic service chaining and are being introduced into geospatial domain. Currently syntactic specification has its advantage in having concrete and industry-wide tools. To take the best from both approaches, this paper provides an approach to automatically transform semantic description from Semantic Web approaches to syntactic specification for geo-processing service chains. Both syntactic specification and semantic description for OGC standards-compliant geospatial Web services and geo-processing chains are provided and illustrated using OWL-S and BPEL. Transformation strategy is presented and a prototype tool transforming OWL-S to BPEL is implemented to demonstrate the applicability of our approach.

Keywords: Geospatial, Geo-Processing, Service Chain, Semantic Web, BPEL, OWL-S, OWLS2BPEL.

1 Introduction

Recently, Service-Oriented Architecture (SOA), as a new information infrastructure, is being introduced into scientific research, such as Earth System Grid (ESG), GEON-Grid and UK e-science program. In 2005, Ian Forster put forward the concept of Service-Oriented Science [1], referring to the scientific research enabled by the SOA. With this information architecture, large volumes of geospatial data and diverse processing functions are available as services for worldwide open use. These services can be chained flexibly to construct different geo-processing workflows. A complex geo-processing service chain, also known as a geo-processing workflow, is scattered among multiple service providers. Therefore, standards for publishing, finding, binding and

M. Bertolotto, C. Ray, and X. Li (Eds.): W2GIS 2008, LNCS 5373, pp. 50–62, 2008.

execution of services are needed. By following the standards of interfaces, the interoperability of different software systems is achieved and Web services developed by different organizations can be combined to fulfill users' requests. In the geospatial Web services area, the Open Geospatial Consortium (OGC) is the leading organization working on developing geospatial Web services standards by adapting or extending the common Web service standards. Through the OGC Web Services (OWS) testbeds, OGC has been developing a series of interface specifications under the OGC Abstract Service Architecture, including Web Feature Service (WFS), Web Map Service (WMS), Web Coverage Service (WCS), Sensor Observation Service (SOS), Catalogue Service for Web (CSW), and Web Processing Service (WPS).

However, current standards focus on syntactic specification and do not address the semantics of geospatial Web services [2]. Semantic Web technologies [3] provide promises for achieving semantic interoperability of geospatial Web services. Semantic Web Services, the combination of Semantic Web and Web Services, aim to provide mechanisms for organizing information and services so that the correct relationships between available data and services can be determined automatically and thus help build workflows dynamically for specific problems. Presently, Semantic Web technologies and in particular Semantic Web Services, have been widely used to enable dynamic service chaining [4][5][6] and are being introduced into geospatial domain for creating geo-processing service chains [7][8][9][10].

In the general information domain, the major international bodies setting the Web service standards are World Wide Web Consortium (W3C) and Organization for the Advancement of Structured Information Standards (OASIS). The Web Services Business Process Execution Language (WSBPEL) [11], shortly known as BPEL, has been approved by OASIS as an industry-wide standard which can be used for syntactic specification of service chains. There is no similar standard available yet in geospatial domain and OGC has adopted BPEL for its interoperability experiments. Although BPEL can not support dynamic service chaining directly as Semantic Web approaches do, it has its advantage in having concrete and industry-wide tools. To take the best from both approaches in syntactic and semantic worlds, this paper provides an approach to automatically transform semantic description from Semantic Web approaches to syntactic specification for geo-processing service chains. OGC standards-compliant services are aligned with the Web service standards from W3C and OASIS to support the syntactic specification in BPEL. Web Ontology Language (OWL) based Web service ontology (OWL-S) [12], a Semantic Web Services technology based on the standard Web ontology language (i.e. OWL) recommended by W3C, is used for the semantic description of geospatial Web services and geo-processing service chains. Finally, a prototype tool transforming OWL-S to BPEL is implemented to demonstrate the applicability of our approach.

2   Geo-Processing Service Chain – An Example

A geospatial application serves as an example which uses distributed heterogeneous data and various geo-processing services. It generates the wildfire prediction data prod service chain illustrated in Figure 1. The input data includes both weather and remote sensing data. National Oceanic & Atmospheric Administration (NOAA) National

Digital Forecast Database (NDFD)[1] provides the daily maximum temperature (MAXT), daily minimum temperature (MINT), and precipitation (QPF) data. National Aeronautics and Space Administration (NASA) Earth Observing System (EOS) [2] provides the Leaf Area Index (LAI), Fraction of Photosynthetically Active Radiation (FPAR), Land Cover/Use Types (LULC) data. These data are accessible through the standard-compliant data providing service, i.e. WCS. General geospatial data processing services including Coordinate Transformation Service (CTS)[3], DFTS (Data Format Transformation Service) and Resolution Conversion Service (RCS), are needed to transform the NDFD and MODIS data into the form that can be readily accepted by the wildfire prediction processing service.

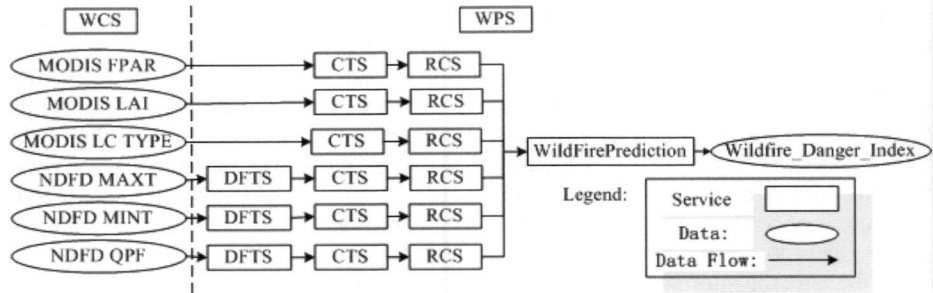

Fig. 1. A geo-processing service chain for the wildfire prediction application

3   Syntactic Specification of Geo-Processing Service Chains

3.1   Syntactic Specification of Geospatial Web Services

Currently, OGC Web services are not equivalent to the W3C SOAP[4]-based Web services [13]. Most of OGC Web service implementations provide access via HTTP GET, HTTP POST and do not support SOAP. This leads to the problem relating to the use of BPEL for syntactic specification of geo-processing chains, namely the lack of W3C WSDL[5] documents for syntactic specification of geospatial services. Conceptually, the OWS also follows the publish-find-bind paradigm in the SOA and has service discovery, description, and binding layers corresponding to UDDI[6], WSDL, and SOAP in the W3C architecture [13]. In addition, OGC is also attempting to integrate the W3C Web services standards into the OWS framework by providing WSDL descriptions for OGC

---

[1] http://www.nws.noaa.gov/ndfd/

[2] http://edcdaac.usgs.gov/datapool/datatypes.asp

[3] OGC Web Coordinate Transformation Service (WCTS) also is treated as a processing service and aligned with WPS by defining input/output WPS complex data type with the schema following WCTS.

[4] Simple Object Access Protocol (SOAP) Version 1.2, http://www.w3.org/TR/2001/WD-soap12-part0-20011217/

[5] Web Services Description Language (WSDL) 1.1. http://www.w3.org/TR/ 2001/NOTE-wsdl-20010315

[6] The Universal Discovery Description and Integration (UDDI) technical white paper, http://uddi.org/pubs/uddi-tech-wp.pdf

Web services [14]. Therefore, in this paper, we rely on the WSDL for the syntactic specification of all geospatial Web service.

The three mandatory operations included in the standard WPS interface are GetCapabilities, DescribeProcess and Execute [15]. Correspondingly, these operations are defined in the WSDL document. For example, in Figure 2, the Execute operation takes an input message with the part defined using an existing element (i.e. wps:Execute in Figure 2) from WPS schemas, and an output message with the part defined using the element "wps:ExecuteResponse" from WPS schemas. The WSDL document can include HTTP GET and POST bindings. Since different processes in WPS may introduce arbitrary parameters, Key-Value Pair (KVP) definitions in the WSDL still need a similar solution as WCS does, i.e. using a comma separated string value for certain parameters [14]. This paper takes the HTTP POST binding using XML document for WPS as the starting point. More technical details please refer to [14].

```
<message name="Execute_POST"><part name="payload" element="wps:Execute"/>
</message>
<message name="ExecuteResponse"><part name="payload"
element="wps:ExecuteResponse"/></message>
<portType name="WPS_HTTP_POST_PortType">
   <operation name="Execute">
      <input message="tns:Execute_POST"/>
      <output message=" tns:ExecuteResponse"/>
   </operation>
   …
</portType>
<binding name="WPS_HTTP_POST_Binding" type=" tns:WPS_HTTP_POST_PortType">
   <http:binding verb="POST"/>
   <operation name="Execute">
      <http:operation location="wps"/>
      <input><mime:mimeXml part="payload"/>
         <mime:content type="application/x-www-form-urlencoded"/>
      </input>
      <output><mime:mimeXml part="payload"/>
         <mime:content type="text/xml"/>
      </output>
   </operation>
   …
</binding>
<service name="GMU-NGA-WPS">
   <port name="WPS_HTTP_POST" binding=" tns:WPS_HTTP_POST_Binding">
      <http:address location="http://geobrain.laits.gmu.edu:8099/languageparser/"/>
   </port>
</service>
```

Fig. 2. A snippet of the WSDL document for a WPS

3.2 Syntactic Specification of Geo-Processing Chains

Figure 3 illustrates the key components in BPEL. An executable BPEL process can provide the process description for a geo-processing service chain based on a number of activities, partners and messages exchanged between these partners. A BPEL process is exposed as a Web service using WSDL. Activities in BPEL are divided into two classes: basic and structured. Basic activities such as "Assign", "Invoke", "Reply" and "Receive" describe elemental steps of the process behavior. Structured activities such as "Sequence" and "Flow" encode control-flow logic, and therefore can contain other basic and/or structured activities recursively. Web services with which a BPEL process interacts are associated as partner links using their accompanying WSDL. Each partner link is characterized by a "partnerLinkType" which contains the role and port type that the role must support for the interaction. The variables defined in BPEL contain messages that constitute a part of the state of a BPEL process and are often passed between partners [11].

```
<process…>
  <partnerLinks>
    <partnerLink    name="client"    partnerLinkType="client:FireCaseProcessLinkType"
myRole="FireCaseProcessProvider"/>
    <partnerLink  name="PL_WCS"  partnerRole="WCS_HTTP_GET_PortTypeProvider"
partnerLinkType="lns:GMU-NGA-WCSLinkType"/>
    <partnerLink  name="PL_WPS"  partnerRole="WPS_HTTP_POST_PortTypeProvider"
partnerLinkType="lns:GMU-NGA-WPSLinkType"/>
  </partnerLinks>
  <variables>
    <variable messageType="wcs:GetCoverage_GET"
name="GetCoverage_GET_request_0"/>
    <variable messageType="wcs:GetCoverageResponse"
name="GetCoverageResponse_request_0"/>
    …
  </variables>
  <sequence>
    <receive …>
    <flow …>
      <sequence …><assign …><copy …>…<invoke …>…
      <sequence …>
      …
    </flow>
    <assign …>
    <reply …>
  </sequence>
</process>
```

Fig. 3. Key components in BPEL

4   Semantic Description of Geo-Processing Service Chains

As shown in Figure 4, OWL-S is structured in three main parts:

(1) Service profile: what a service does (advertisement).

(2) Service model: how a service works (detailed description), e.g., a series of necessary input parameters identified in the service model. A service model can be either a "Simple", "Atomic" or "Composite" process.

(3) Service grounding: how to assess a service (execution), e.g., grounding the input/output ontology concepts to the output message of WSDL operation using Extensible Stylesheet Language Transformations (XSLT).

In [7], we have defined the geospatial "DataType" and "ServiceType" ontologies to address the data and functional semantics of services respectively. Data semantics annotate the semantics of input and output data in a Web service operation. Functional semantics represent the semantics for a service function. The "DataType" ontology, in the service grounding, can be used to define a set of bidirectional mappings between the schemas of the OGC-compliant services and the ontologies. "Atomic Process" ontology in OWL-S is used for semantic description of geospatial Web services, while "Composite Process" ontology in OWL-S is used for semantic description of geo-processing service chains.

```
<!--snippet of an OWL-S-->
<!-- Service description -->
<service:Service rdf:ID="wildfireprediction_service_01">
    <service:describedBy rdf:resource="#wildfireprediction_process_01"/>
    <service:presents rdf:resource="#wildfireprediction_profile_01"/>
    <service:supports rdf:resource="#wildfireprediction_wsdlgrounding_01"/>
</service:Service>
<!-- Profile description -->
<profile:Profile rdf:ID="wildfireprediction_profile_01">...</profile:Profile>
<!-- Process Model description -->
<process:AtomicProcess rdf:ID="wildfireprediction_process_01"> ...
</process:AtomicProcess>
<process:Input rdf:ID="wildfireprediction_input_lai">
    <process:parameterType rdf:datatype="&xsd;#anyURI">
    &geodatatype;#LAI</process:parameterType>
</process:Input>
<!-- Grounding description -->
<grounding:WsdlGrounding rdf:ID="wildfireprediction_wsdlgrounding_01">
    <grounding:hasAtomicProcessGrounding
rdf:resource="#wildfireprediction_wsdlatomicprocessgrounding_laits01"/>
</grounding:WsdlGrounding>

<!--snippet of a composite process-->
<process:CompositeProcess ...>
<process:composedOf>
    <process:Sequence>
     <process:components>
      <process:ControlConstructList>
       <list:first>
        <process:Split-Join>
         <process:components>
             <process:ControlConstructBag>...
```

Fig. 4. Key components in OWL-S

5   Automatic Transformation

This paper focuses on generating an executable BPEL process from OWL-S. The BPEL abstract process and OWL-S simple process are not supported since both of them are abstract views of processes and not invocable. Therefore, the following transformation strategy applies to OWL-S atomic and composite processes only. The starting point of transformation is the interface description (i.e. WSDL document) for the resulting BPEL process, because the transformation result is exposed as a Web service and the description of the BPEL process refers to the elements in its WSDL document.

5.1   Generating WSDL Document for BPEL

Figure 5 gives an overview of the process for generating WSDL document for BPEL. An OWL-S atomic process is a description of a service that expects one message and returns one message in response [12]. The WSDL document of a BPEL process generated from an OWL-S atomic process is simple. The messages can refer to definitions from the accompanying WSDL in the WSDL grounding of the atomic process. An operation similar to the accompanying WSDL in the WSDL grounding can be defined based on the input and output messages, and a new port type is described based on this operation. Two partner link types are generated automatically: one is for the participating Web service and the other one is for the exposed BPEL Web service itself.

An OWL-S composite process specifies the interaction among multiple Web services. Therefore, the difference between the BPEL WSDL generation of the composite process and that of atomic process is that in the transformation process for an OWL-S composite process, the input/output message is the union of input/output messages for those Web services at the beginning/end of the composite process, and the partner link types includes all participating Web services. All parts of each input/output message for those services at the beginning/end of a composite process are

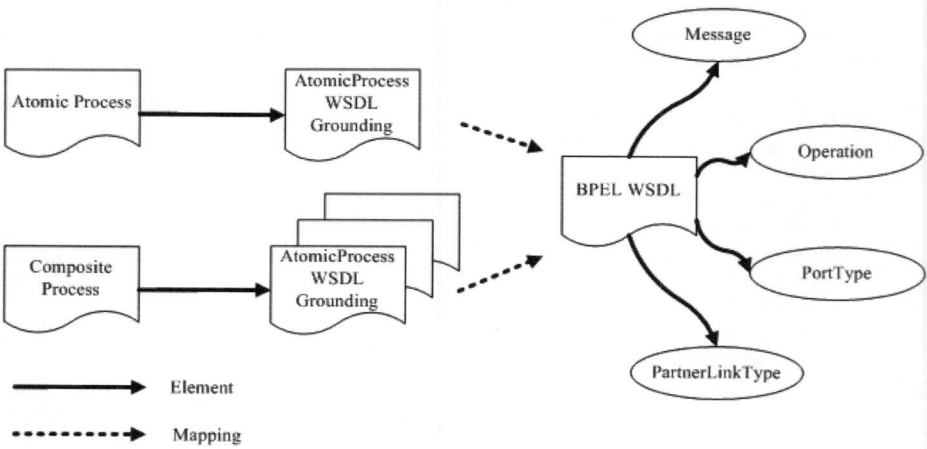

Fig. 5. Overview of the process for generating WSDL document for BPEL

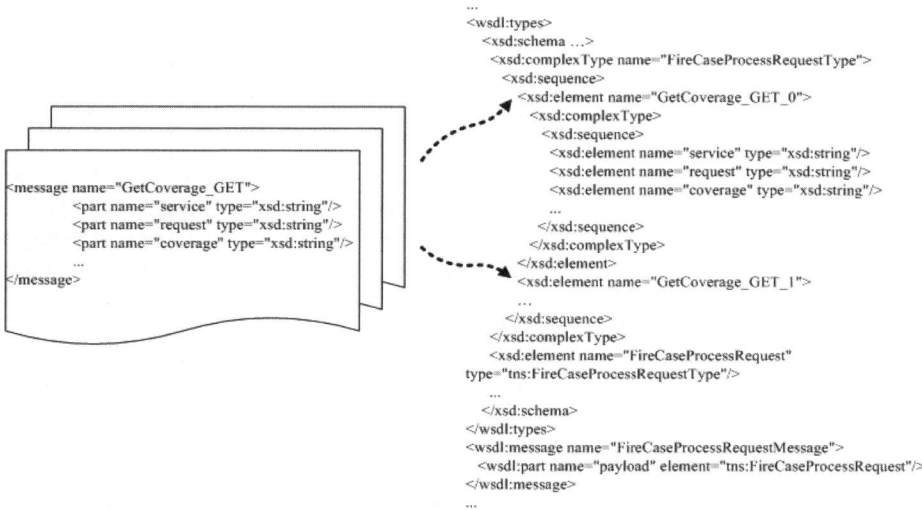

Fig. 6. Messages union strategy for BPEL WSDL document

converted into a child complex type of the schema type for input/output message of the BPEL WSDL. The element name for this child complex type has its prefix from the original message to help identify the data flow mentioned later in section 5.2. Figure 6 illustrates our strategy for messages union. Taking the use case in section 2 as an example, totally six child complex types from WCS requests are defined in the request type of the generated WSDL document. Three partner link types are defined as shown in Figure 7: the first one is for WCS, the second one is for WPS and finally the third one is for BPEL service itself.

```
<plnk:partnerLinkType name="GMU-NGA-WCSLinkType">
   <plnk:role name="WCS_HTTP_GET_PortTypeProvider">
      <plnk:portType name="wcs:WCS_HTTP_GET_PortType"/>
   </plnk:role>
</plnk:partnerLinkType>
<plnk:partnerLinkType name="GMU-NGA-WPSLinkType">
   <plnk:role name="WPS_HTTP_POST_PortTypeProvider">
      <plnk:portType name="wps:WPS_HTTP_POST_PortType"/>
   </plnk:role>
</plnk:partnerLinkType>
<plnk:partnerLinkType name="FireCaseProcessLinkType">
   <plnk:role name="FireCaseProcessProvider">
      <plnk:portType name="tns:FireCaseProcess"/>
   </plnk:role>
</plnk:partnerLinkType>
```

Fig. 7. Partner link type definitions in the WSDL document for the wildfire prediction case

5.2   Generating BPEL from OWL-S

Figure 8 gives an overview of the process for generating BPEL from OWL-S composite process. The generation of a BPEL process for an OWL-S atomic process is just a simplified version of transformation process for OWL-S composite process. Therefore, we describe details of the transformation process only for the composite process. A composite process can be characterized as a collection of subprocesses with control and data flow relationships. In OWL-S, the control flow is represented by the control constructs such as Sequence and Split. The data flow is specified through input/output bindings using a class such as ValueOf to state that the input to one subprocess should be the output of the previous one.

The transformation process follows the control flow in the OWL-S composite process. As shown in Figure 8, the control constructs are mapped into structured activities in BPEL. Meanwhile, partner links are generated for partner link types in the BPEL WSDL document and each pair of input and output variables is generated according to the "wsdlOperation" in the grounding of each atomic process. "Perform" in OWL-S indicates an invocation of a process, therefore an "Invoke" activity is generated for each "Perform" when transversing the control flow. The "receive" and "reply" activities specify services that a BPEL process provided to its partners, thus they are generated respectively at the beginning and end of the transformation for the control flow.

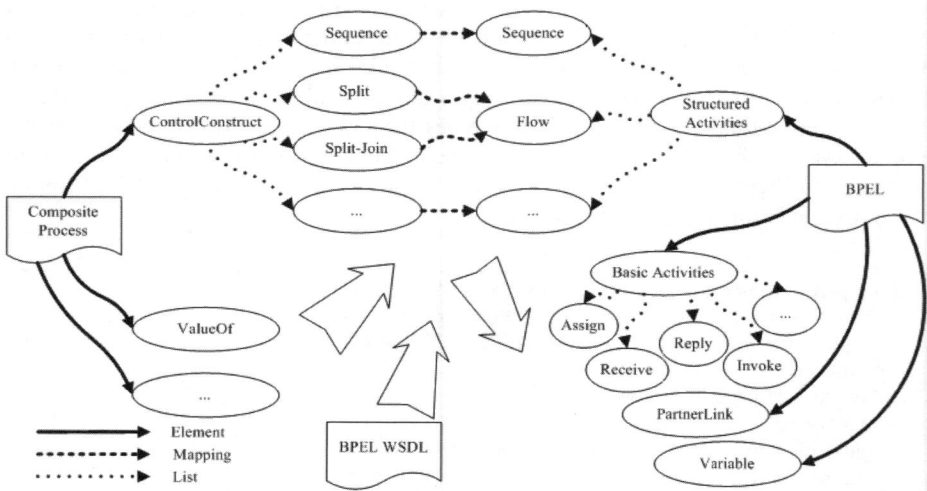

Fig. 8. Overview of the process for generating BPEL from OWL-S composite process

The data flow in OWL-S makes the process of generating "copy" elements in the "assign" activity a non-trial issue during the transformation process. For example, the "ValueOf" specifies the data flow at an abstract level, i.e. the binding parameters are conceptual types in the semantic world. Therefore, it must be combined with the grounding information to build the syntactic mapping between the message parts of dependable services. As we pointed in [7], a mediated RDF structure is necessary for the service grounding description in OWL-S; it serves as the relay structure from the

XML structure of the output of one service to the input XML structure of the next service. Message mappings between services are indirectly embedded in the mapping of the services message schema structure to a mediated RDF structure in the service grounding of OWL-S. Therefore, we rely on this mediated RDF structure to generate the "copy" elements in the BPEL.

5.3 Supporting OWS Complex Types

Currently, OWL-S is still under development. Our application identifies that the WSDL grounding in OWL-S cannot well handle the mapping of the multiple OWL-S inputs to a complex schema type (e.g. wps:Execute in WPS schema) in the message, while in OWS, XML schema type is often used and offers the flexibility in terms of the complexity of OWS message. In addition, in the current WPS specification [16], KVP encoding for "Execute" operation is optional while XML encoding is mandatory. Following the approach of XML Message handlings in BPEL, the WSDL grounding is extended by an additional property "wsdlMessagePartElement" which contains the XPATH[7] to locate the certain element in the complex type. Table 1 shows a snippet of WSDL grounding and corresponding BPEL for a WPS process. A variable is generated for each message part element which has a complex schema

Table 1. A snippet of WSDL grounding and corresponding BPEL for a WPS process

WSDL Grounding	<grounding:WsdlInputMessageMap ...>
	<grounding:owlsParameter rdf:resource="#wildfireprediction_input_lai"/>
	<grounding:wsdlMessagePart ...>&wildfireprediction_wsdl;#payload</grounding:wsdlMessagePart> ...
	<groundingx:wsdlMessagePartElement
	rdf:datatype="&xsd;#string"><![CDATA[<context type="xpath"
	xmlns="http://www.laits.gmu.edu/geo/ontology/domain/groundingx.owl"
	xmlns:wps="http://www.opengeospatial.net/wps"
	xmlns:xlink="http://www.w3.org/1999/xlink">**/wps:Execute/wps:DataInputs/wps:Input[position()=5]/wps:LiteralValue** </context>
	]]></groundingx:wsdlMessagePartElement>
	<grounding:xsltTransformationString>...

BPEL	<variable name="Execute_fire_wps" element="wps:Execute"/> <variable name="Fire_ InputVariable" messageType="wps:Execute_POST"/> ... <copy> <from variable="lai_wps_resampling_OutputVariable" part="payload" query="/wps:ExecuteResponse/wps:ProcessOutputs/wps:Output[position()=5]/wps:ComplexValue/wcts:TransformedData/wcts:Reference/@xlink:href"/> <to variable="Execute_fire_wps" query="**/wps:Execute/wps:DataInputs/wps:Input[position()=5]/wps:LiteralValue**"/> </copy> <copy> <from variable="Execute_fire_wps"/> <to variable="Fire_ InputVariable" part="payload" query="/wps:Execute"/> </copy>...

---

[7] XML Path Language. W3C, http://www.w3.org/TR/xpath

type. The XPATH is extracted from the property "wsdlMessagePartElement" and assigned to a child node (i.e. "to" element) of a "copy" element, where the mapping relation between "from" and "to" elements is determined by the transformation strategy mentioned in the last paragraph of section 5.2.

6 Implementation

For the implementation of the proposed transformation strategy, we use existing tools for both OWL-S and BPEL. OWL-S API[8] provides a Java API for programmatic access to read, execute and write OWL-S service descriptions. The API provides an ExecutionEngine that can invoke AtomicProcesses that has WSDL groundings, and CompositeProcecesses that uses control constructs such as Sequence, and Split-Join. It is further extended in this work to support the HTTP GET and POST invocation in addition to its just SOAP invocation. The support to "wsdlMessagePartElement" is also added in this open source software. ActiveBPEL[9] is used for writing BPEL and BPEL WSDL descriptions. It is an open source BPEL library which provides full support to all components of BPEL.

An OWL-S to BPEL conversion tool has been developed and implemented. It works as a Web application[10]. The conversion results can be sent to a BPEL engine[11] for execution. The conversion tool supports the HTTP GET binding for WCS and SOAP-based Web services for geo-processing services. At the time to develop this tool, WPS was still not known to a wider group. Since it is now emerging as a forthcoming standard specification, we already took an initial effort to complement the WSDL grounding in OWL-S and the BPEL for WPS in the JDeveloper BPEL Designer from Oracle BPEL Process Manage 10.1.2, which is a robust designer for BPEL and can provide much convenience for testing conversion results. A future goal is to aggregate the current support for WPS into the existing conversion tool.

7 Related Work and Discussion

There have been studies focusing on mapping BPEL to OWL-S [16][17] in the general information domain. However, as pointed out in their work, the derived mapping specification in this direction lacks semantic concepts that facilitate the composition of services when using OWL-S. In other words, this approach takes a bottom-up approach to connect the syntactic world to the semantic world. From this perspective, our approach takes a top-down approach since it grounds concepts in the semantic world into the descriptions in the syntactic world.

It is noted that existing work in the general information domain [16] [17] has identified incomplete mappings between BPEL and OWL-S such as some control flow

---

[8] http://www.mindswap.org/2004/owl-s/api/
[9] http://www.active-endpoints.com/
active-bpel-engine-overview.htm
[10] http://www.laits.gmu.edu:8099/OWLS2BPEL/
[11] A BPEL engine developed by LAITS, GMU, available at
http://geobrain.laits.gmu.edu/

pattern like choice, mainly because OWL-S is still under development and not as much mature as BPEL. The same limitation exists in our work. The focus of our work is not trying to provide a complete solution to solve these incomplete mappings, rather, inspired by the requirement in the geospatial domain, we provide an initial effort towards the mapping in an opposite direction. On the one hand, the application in the geospatial domain can build its work on the existing efforts in the general information domain towards the mapping specification. On the other hand, this application-driven effort would complement the research in the general information domain.

To ensure correctness of syntactic specifications in BPEL, the transformation relies heavily on the grounding information in OWL-S. The description for service groundings is usually complex and error-prone in sophisticated business processes. However, focusing on the OGC-compliant geospatial services only, it is possible to define some common OWL-S grounding representations for all OGC service instances under the same standard interface and message as illustrated in [7]. And this would greatly support the wide application of our approach.

8 Conclusions

Although semantic Web technologies have shown a promising prospect for supporting semantic interoperability and dynamic service chaining in geospatial domain, syntactic specification presently has its advantage in having concrete and industry-wide tools. This paper provides an approach to take the best of both by automatically generating syntactic specification from semantic description for geo-processing service chains. The effort of this paper can not only facilitate the wider application of semantic Web technologies in geospatial domain, but also provide a valuable input to the research in the general information domain.

Acknowledgements

This work is supported by U.S. National Geospatial-Intelligence Agency NURI program (HM1582-04-1-2021), Project 40801153 supported by NSFC, National High Technology Research and Development Program of China (863 Program, No. 2007AA12Z214) and Specialized Research Fund for State Key Laboratory of Information Engineering in Surveying, Mapping and Remote Sensing of China.

References

1. Forster, I.: Service-oriented science. Science 308(5723), 814–817 (2005)
2. Percivall, G. (ed.): The OpenGIS Abstact Specification, Topic 12: OpenGIS Service Architecture, Version 4.3, OGC 02-112. Open GIS Consortium Inc. 78 pages (2002)
3. Berners-Lee, T., Hendler, J., Lassila, O.: The semantic web. Scientific American 284(5), 34–43 (2001)
4. Srivastava, B., Koehler, J.: Web service composition - current solutions and open problems. In: Proceedings of ICAPS 2003 Workshop on Planning for Web Services, Trento, Italy, pp. 28–35 (2003)

5. Rao, J., Su, X.: A survey of automated web service composition methods. In: Cardoso, J., Sheth, A.P. (eds.) SWSWPC 2004. LNCS, vol. 3387, pp. 43–54. Springer, Heidelberg (2005)
6. Peer, J.: Web service composition as AI planning - a survey. 63 pages, University of St.Gallen, Switzerland (2005)
7. Yue, P., Di, L., Yang, W., Yu, G., Zhao, P.: Semantics-based automatic composition of geospatial Web services chains. Computers & Geosciences 33(5), 649–665 (2007)
8. Lutz, M.: Ontology-based descriptions for semantic discovery and composition of geo-processing services. Geoinformatica 11(1), 1–36 (2007)
9. Lemmens, R., Wytzisk, A.: Integrating semantic and syntactic descriptions to chain geo-graphic services. IEEE Internet Computing 10(5), 18–28 (2006)
10. Roman, D., Klien, E., Skogan, D.: SWING - A semantic web services framework for the geospatial domain, Terra Cognita 2006. In: International Semantic Web Conference ISWC 2006 Workshop, Athens, Georgia, 4 pages (2006)
11. OASIS: Web services business process execution language, version 2.0. Web Services Business Process Execution Language(WSBPEL) Technical Committee(TC), 264 pages (2007)
12. Martin, D., et al.: OWL-based web service ontology (OWL-S) (2006),
    `http://www.daml.org/services/owl-s`
13. Tu, S., Abdelguerfi, M.: Web services for geographic information systems. IEEE Internet Computing 10(5), 13–15 (2006)
14. Sonnet, J. (ed.): OWS 2 common architecture: WSDL SOAP UDDI. Version: 1.0.0. OGC 04-060r1. Open Geospatial Consortium, Inc., 76 pages (2005),
    `https://portal.opengeospatial.org/files/?artifact_id=8348`
15. Schut, P., Whiteside, A.: OpengGIS® web processing service, Version 0.4.0, OGC 05-007r4, Open Geospatial Consortium, Inc., 86pages (2005)
16. Shen, J., Yang, Y., Zhu, C., Wan, C.: From BPEL4WS to OWL-S: integrating e-business process descriptions. In: The 2005 IEEE International Conference on Services Computing (SCC 2005), Orlando, FL, pp. 181–188 (2005)
17. Aslam, M.A., Auer, S., Shen, J.: From BPEL4WS process model to full OWL-S ontology. In: Proceedings of Posters and Demos 3rd European Semantic Web Conference (ESWC 2006), Budva, Montenegro, pp. 61–62 (2006)

Online Generation and Dissemination of Disaster Information Based on Satellite Remote Sensing Data

Chao-Yang Fang[1], Hui Lin[1], Qing Xu[1], Dan-Ling Tang[2],
Wong-Chiu Anthony Wang[1], Jun-Xian Zhang[1], and Yick-Cheung Matthew Pang[1]

[1] Institute of Space and Earth Information Science, The Chinese University of Hong Kong,
Shatin, N. T., Hong Kong
{fangchaoyang,huilin,xuqing,chiuwang,zhangjunxian,
Pangyc}@cuhk.edu.hk
[2] Laboratory for Tropical Marine Environmental Dynamics, South China Sea Institute of
Oceanology, Chinese Academy of Sciences, Guangzhou, China
lingzistdl@126.com

Abstract. This paper aims to introduce a system framework for online generation and dissemination of disaster information based on satellite remote sensing
data, that is, the design and application of the WEB-based Disaster Monitoring
and Warning Information System – WEB-DMWIS. By integrating the satellite
remote sensing data, scientific workflow technique, and the Application Service
Provider (ASP) service model, WEB-DMWIS makes it possible for different
levels of users to access customized disaster information via WEB. In this way,
it provides an information platform which supports widely-used near real-time
monitoring and early warning of natural disasters. The system is composed of
two parts: (a) a user-side browser; (b) a server-side disaster information center.
After introducing the system framework and functions of main engines, this paper described in detail the principle and realization of the online generation and
distribution of disaster information. In addition, the disaster monitoring and
early warning information system for fishery industry in Pearl River Delta region is developed on the basis of the WEB-DMWIS to verify and demonstrate
the effectiveness of the framework.

Keywords: Disaster Monitoring, Online Generation, Online Dissemination,
Satellite Remote Sensing.

1 Introduction

During the last 10 years in last century, a large number of satellites carrying multiple
sensors (visible, infrared and microwave) have been launched and applied, which
makes it possible to monitor the land, the ocean, the atmosphere, and the ionosphere
day and night, and rain or shine[4]. The data from these satellites on natural environment, economic development and human activity have played a great potential in the
monitoring and mapping of loss due to various natural disasters (earthquakes, land-
slides, floods, volcanoes, cyclone / hurricanes, harmful algal blooms, water quality,
oil spills, dust storms, droughts, etc.)[5, 8]. Satellite remote sensing data have also

M. Bertolotto, C. Ray, and X. Li (Eds.): W2GIS 2008, LNCS 5373, pp. 63–74, 2008.
© Springer-Verlag Berlin Heidelberg 2008

been used to analyze and understand the reasons and impacts of the disasters. In a word, satellite remote sensing data have become the main data base for the monitoring and early warning system of natural disasters thanks to their large spatial coverage, relatively quick data access and update, long-term time series and relatively high spatial resolution. It can be said that at present remote sensing data have been hardly unused in the process of monitoring and assessment of any serious natural disasters [13]. Natural disasters usually happen suddenly and have extensive impacts. Therefore, the speed and interactivity of the system must be taken into consideration when a natural disaster monitoring, early warning and decision-supporting system is developed, in order to generate the disaster information in a prompt and timely manner as well as to disseminate the information to users at different levels for their convenience. It is clear that the WEB is the supporting platform which meets these requirements and has been generally accepted and widely used [9, 15 and 19]. It has exercised immense influence in our work and life. In fact, with the development of IT and demonstration system, plus the double promotion of technology and demand, the WEB technology is now becoming more and more mature and widely accepted in the management and application of satellite remote sensing data [7, 16, 17 and 20].

However, there are few systems which can realize the online generation and distribution of disaster information based on satellite remote sensing data. The WEB has been playing a significant role in query and access to satellite remote sensing data. The major centers of satellite data processing and distribution in the world have set up their own data distribution websites [10, 11 and 14]. Through these websites, users can select data by interactive parameter setting and data preview images. In addition to providing Level 1 products, these websites also focus on the processing of some Level 2 and Level 3 products which contain some thematic information for users to download. Furthermore, they will also release some detailed pictures of some most serious disasters (such as oil spills and fires) [12, 13] with some analysis and interpretation. However, to generate and dissemination these disaster information, researchers usually use remote sensing data processing and mapping system in an environment without WEB system and then post images about the historical disasters on web with some specific interpretation.

In respect of disaster information visualization, the users' interaction is increasingly supported. Not only the users can preview the image when querying the data, but they can interactively customize the required expression forms. However, the current commercial WEBGIS is unable to provide support because of the complexity of the disaster information retrieval processes and many interactive controls they take. So in the three modules of data management, analysis and visualization, the interactive support of data analysis is hardly realized on the basis of the present network. However, different users care about different space, time and content levels. The visualization forms they require are also different. In this sense, the pre-produced disaster information products might fail to meet specific requirements. Therefore, for the establishment of natural disaster monitoring and early warning information system, it is very important to design and develop a WEBGIS framework which supports users to interactively order the customized disaster information on the WEB-based platform.

In this paper, the design and application of a WEB-based Disaster Monitoring and Warning Information System (WEB-DMWIS)—a system framework-- will be introduced for online generation and distribution of disaster information based on satellite remote sensing data. By integrating the satellite remote sensing data, scientific workflow technique, and the Application Service Provider (ASP) service model, WEB-DMWIS makes it possible for different levels of users to access customized disaster information via WEB. In this way the near real-time monitoring and early warning of natural disasters can be realized. This paper is organized as follows: an overview is presented in Section 2 on WEB-DMWIS including user demands analysis, process of online generation, and distribution of disaster information; the design and implementation of the framework is given in Section 3. Section 4 is a case study of its application. Conclusions and the direction of future development are discussed in Section 5.

2   Overview of WEB-DMWIS

2.1   User Demand's Analysis

With the design and development of near-real-time disaster monitoring and early warning information system which is based on WEB technology and satellite remote sensing data, we aim to start the corresponding disaster information extraction process and generate online customized disaster information according to the customized requests of different levels of users. The system will also map and render images in accordance with the users' requirements, and then dissemination them in near real time to users through the web service. Different users usually care about different disaster information such as disaster information in terms of time, or position, or details? or temporal and spatial scales, or presentation format. Thus, how to describe and organize the requirements of these users and realize the efficient management is one of the most important issues in the process of design and development of the system. In addition, the traditional methodology which completes the information extraction beforehand and then provides pre-processed images for users to download can not meet the customized requirements of users. Therefore, the corresponding disaster information extraction processes in accordance with the users' requirements need to be triggered for online generation and distribution of disaster information. These processes consist of satellite remote sensing data which contain disaster information and a series of necessary data processing, analysis and expression methods to extract and mine information from these data. Managing and scheduling the data and methods so that the disaster information extraction processes can be carried out correctly and fluently are the other two key issues. To conclude, the operation of the whole system is arranged and organized by user requirement management engine, data management engine and method management engine.

In the aspect of remote sensing principle, satellite data used in natural disaster information extraction include not only visible light and infrared data but also microwave data. In the aspect of data level, level 1, level 2 and level 3 data products can all be considered [6]. In addition, different data analysis and visualization methods will have different requirements on data content and data format in the process of information extraction. Therefore, the data management engine should perform the following functions:

- Import and manage various satellite remote sensing data with different spatial resolutions and sources. It is also responsible for data pre-processing, quality control, registration and metadata management. A data model which generalizes all capabilities of various data will be applied to realize the integrated management of data.
- Enquire and gather data required to extract disaster information according to requirements submitted by users. It will then process and combine the data into suitable contents and formats so that they can be provided to information extraction processes.
- Organize and manage the intermediate data generated during disaster information processing. It guarantees the generation of disaster information and automated image rendering.
- Dissemination and manage the final disaster information and images and return them to the users. At the same time, it will update the disaster information database and archive the generated knowledge to history records so that the other users can call them quickly.

Extracting disaster information from satellite remote sensing data is a complex and professional process. Disaster information extraction methodology and modeling have been studied for many years and fruitful results have been obtained. However, most of these methods and models were released in scientific papers and technical reports. Thus, these research results are in a discrete, isolated and static status which can not be efficiently organized into the processes that can be automatically operated [22]. In our system the technology of scientific workflow is introduced. With the establishment of the scientific workflow engine, automatic extraction of disaster information can be realized [1, 2 and 21]. This engine performs the following functions:

- Implement basic functions, such as data preprocessing, analysis and visualization, and modulization of specific information extraction models. It packs the discrete methods and models into functional units which can be used to assemble the disaster information extraction processes.
- Create the workflows of disaster information extraction. It also provides functions such as modifying, editing, saving and loading functions for the workflows.
- Start, monitor and complete the operation of disaster information extraction workflows.
- Query and preview the workflow of disaster information extraction.

The user requirement management engine is the interface between the client-end users and the disaster information service centre. With the management of data flows and control flows, it is responsible to collect users' requirements, staring disaster information extraction process and returning the information at the same time. The user requirement engine established by means of ASP in our system has the following functions:

- Accept the requirements submitted by users, and record their status and submitted services. In this way, the manual for specific tasks can be constructed. It then creates the disaster information extraction task and passes it to the scientific workflow engine for further operation.
- Manage all kinds of status signs during the operation of disaster information extraction so that the interaction and coordination between users and disaster information service centre will be put under control.
- Coordinate and manage various information extraction tasks submitted by different users in the disaster information service centre. Reallocate the resources rationally in order to improve the efficiency.
- Archive and manage the user information and users' queries for statistical analysis of users' usage information so that the services of information center can be adjusted.

2.2 Online Generation and Dissemination of Customized Disaster Information

Supported by the three engines mentioned above, the process of online generation and distribution of customized disaster information in the web-based system contain five steps:

1. Users define the needed services for data, analysis and expression format in an HTML form, and then submit the form through the Internet.
2. The user requirement management engine will search the file history and check whether there are disaster information and image file which meet the user's requirements. If such file is found, it will be used as the result and returned to the user. Otherwise, the data management engine will be arranged to search if there are satellite data matching the temporal and spatial requirements. If such data exist, the user task will be constructed and disaster information extraction process will be triggered.
3. The scientific workflow engine will select the satellite remote sensing data which satisfy the temporal requirement in a specific area according to the disaster information requirements. The selected data will be preprocessed and under quality control. Then relative modules will be called to complete the disaster information analysis and extraction. According to the user requirements in disaster information visualization, the disaster information images will be rendered. Later, the user requirement management engine will be notified for the organization and distribution of the disaster information images.
4. The user requirement management engine will return the disaster information and relative data to the client-end browser. These data will be integrated with the base image and then the custom-built information images can be generated. At the same time, the newly generated disaster information will be saved in form of history data files. User activities will also be recorded in order to analyze the user requirement pattern and thus improve the service quality.
5. Users will browse the disaster information and evaluate the degree of satisfaction with that information. If the result information is satisfying, the users can apply for the acquisition of custom-built disaster information report which contains more detailed information. If the users want to change the interested temporal and spatial

information or visualization format, they can change or tune the settings interactively in the browser. Then the user requirement management engine in the server will reorganize the online generation and distribution of disaster information.

3   Design and Implementation of the WEB-DMWIS

3.1 Architecture of the WEB-DMWIS

In order to realize disaster information online generation and dissemination, we have designed and developed the WEB-DMWIS system as shown in Figure 1. The system is designed into a four-tier architecture:

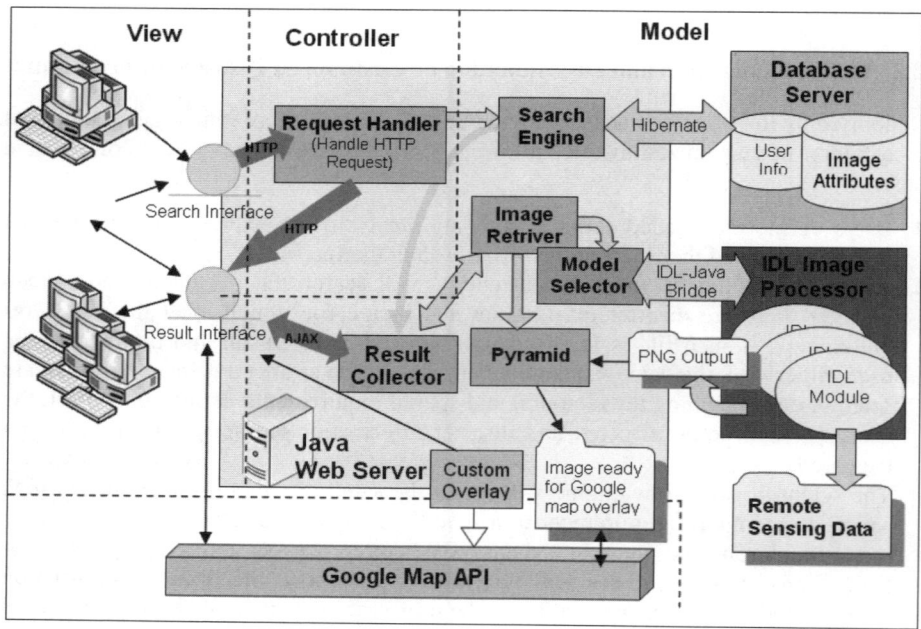

Fig. 1. System Architecture of the WEB-DMWIS

- **View** – the client side which interacts with users directly. It provides interface for parameter inputs and result distribution.
- **Model** – a set of operations that manage the access of the back-end database structure, data collections and other black-box functions with respect to the system. It also handles the object states of the application. Views will send query to models. Models will also notify them when there are object states changes so that corresponding action can be taken in views.
- **Controller** – central collection of all possible queries from the views and choose correct handlers for them. After processing, the controller will dispatch the result to the correct views so that it completes the data flows logically.

- **Google Map API** – a third-party service providing the map service and custom overlay functions. These services enrich the interface in both visualization and manipulation.

3.2 Role of Functional Modules

In this session, the functions in view, model and controller will be discussed.

1) View

Search Interface – collects searching parameters (such as collection date of the data) from users.

Result Interface – shows the searching results by interacting with result collector and Google Map API. It also collects additional image processing parameters.

2) Controller

Request Handler – collects parameters from views. It maintains a predefined list of application behaviors. Such behavior is a specification of which model should be taken to handle a query and which view should be fetched to users after the model finishes processing. Request handler will gather all parameters from the view and pass them to the appropriate model. The model will return its finishing status (if any) to the request handler and the appropriate view will be selected to show the query result to users.

3) Model

- **Search Engine** – all data searching logics are implemented in this model. It will access the database server to retrieve image attributes so that searching can be performed according to the parameters. Searching result will be saved into the "result collector" model. Search engine will notify the request handler whether the search is successful or not.
- **Result Collector** – a special request handling model which handles image processing queries from the result interface. It passes parameters to image processing models and returns the processing result to the view. It maintains a list of searching results.
- **Image Getter** – responsible for monitoring the processing of data in result collector's list according to additional processing parameters gathered from the result interface. Module selector is an auxiliary function which helps image getter to pick up the correct IDL module according to the image processing parameter. Image pyramid will also be arranged in this model. It returns the image attributes, which include the coordinates of north-east corner and south-west corner, number of layers of the pyramid and image path, to the result interface.
- **IDL Image Processor** – a collection of black box functions with respect to the web application which can improve the efficiency in image processing. The IDL module will save the processed images into the location specified by the image getter.
- **Pyramid** – processed image sets in different resolutions. The image generated by the IDL module is too large in size and impractical to be shown in the result interface. Thus, images will be broken down into small pieces when a higher-resolution image is displayed. Pyramid tiles will be saved under the folder which can be accessed by the Internet as well as the Google Map API. This folder location will be returned to the image getter.

3.3 Communication Technologies

In order to realize disaster information online generation and dissemination, four communication technologies have been used in this system to smooth operation of data flow and control flow between the various functional modules. These technologies include:

- **HTTP** – a common communication protocol used between client and web server. It includes client request and server response. Simple query will send HTTP requests or responses through the interfaces.
- **AJAX** – allows web browsers to establish "2-way communication" with the server side and use the server response immediately without fetching to another page. Thus, for example, when the "result collector" returns the image attributes to the result interface, they can be put into the result interface directly instead of opening another interface to show the data.
- **Hibernate** – a Java framework which allows high performance to access the database server by simplifying the model design of database objects.
- **Java-IDL Bridge Connectivity** – a special communication channel established between Java and IDL applications. It is supported by the IDL Connectivity Bridge Library which can export a Java program. This program can call the IDL modules with the functions provided by the Bridge Library's API.

In the next section, we will give a real example of data flow and control flow in the system in application. Because of content limits, the detail of principle and implementation of data model and scientific work flow will be presented in other articles.

4 DMWIS-FIPRD: Disaster Monitoring and Warning Information System for the Fishery Industry in the Peal River Region

Since the early 1980s, the environmental issues have become increasingly prominent with the rapid development of industrialization and urbanization of the Pearl River Delta region. The rivers including the Pearl River are serious polluted. Rivers are seriously polluted, and their water eventually flows into the ocean, which leads to marine pollution. Pearl River is one of them. In its watershed, oil spills caused by well-developed maritime activities such as maritime transport and oil exploitation at sea usually bring large areas of oil pollution, which causes massive death of marine life and leads to enormous economic losses [3]. In order to prevent and reduce the hazards of the harmful algal bloom, water pollution and oil spills on the sea to fisheries and reduce the losses caused by the disasters, we must further enhance marine fisheries and environment monitoring capabilities, as well as major marine environmental disaster monitoring and early warning capabilities [7]. Therefore, a near real-time marine disaster monitoring information system has been developed to monitor the harmful algal bloom, marine environmental pollution and marine oil spills in Pearl River Delta region. The system is designed and developed on the basis of the WEB-DMWIS framework described in Section 2, which consists of two parts, one is browser and the other is server-side marine disaster information center. In our system, the data for operational performance include Moderate Resolution Imaging Spectroradiometer (MODIS) data which contain information of water quality and harmful algal bloom and Advanced Synthetic Aperture Radar (ASAR) data which can be used to detect the oil spills.

MODIS data from the Terra and Aqua satellites of NOAA are received and pre-processed by the Hong Kong Observatory. We are given access to them via FTP within two hours after the satellites pass the region. ASAR data from ESA ENVISAT satellite are received and processed by the satellite ground receiving station at the Chinese University of Hong Kong. In general, the data can be accessed within one hour after the satellite passes the region. The short time delay provides a guarantee for the system's near-real-time disaster monitoring and early warning. To develop the satellite remote sensing data based disaster information monitoring and early warning model, we collected the data of harmful algal bloom and oil spills in the last five years. Besides, 2 in situ measurements were also carried out in the area. Based on these direct and accurate in situ data, the algorithm for water quality parameter extraction, as well as the model for monitoring and early warning of harmful algal bloom and oil spills was developed. The water quality parameters include CDOM (Colored Dissolved Organic Matter), chl-a, TSM (Total Suspended Matter) and CPI (Construction Pollution Index). Similar to the process of generating spatial distribution image of TSM which is shown in Fig. 4, each disaster information extraction process is composed of a series of different functions including data pre-processing, quality control, geometric correction, atmospheric correction, sea and land mask, band algorithm, edge detection, object classification, etc. These processes in the system are organized and operated in form of scientific work flow. The management of scientific work flow is realized by Kepler system in the form of XML file. Kepler is an open-source management system which specifically aims at establishing, operating and verifying the scientific process. For TSM generation process, the establishment and operation of scientific work flow mainly includes the following four main steps:

1. Summarize and parameterize the input and output data for each basic function unit which does data processing, analysis and visualization and for specialized disaster information extraction model. Build the basic function modules of Kepler based on the code realization of the function and model;

2. Build the platform to construct and describe TSM generation process using the visualization flow of Kepler. First of all, establish the information extraction and image mapping and rending, respectively. Then mix these two parts so as to get the disaster information extraction process, which is described and saved in the form of XML file (each XML file includes mission description, component description, flow description and topological relations description.);

3. When a user puts forward the requirement for TSM information generation, as described earlier, while the system runs to interact with IDL (Fig. 2), it actually calls the XML file to implement the corresponding operation, and thus completes information extraction and mapping;

4. When the user gets the information map, he can interactively set up the data stretching parameters and palette if he is not satisfied with the results. According to the user's demand, the management engine will recall the image mapping and rendering module. After that, new maps will be generated online based on the disaster data.

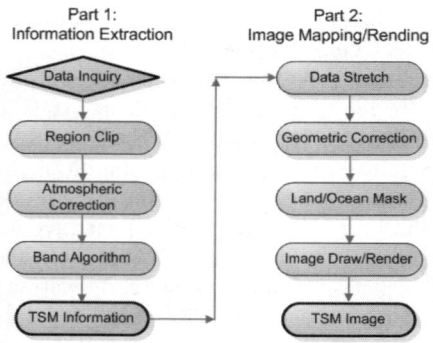

Fig. 2. Process of Generating Spatial Distribution Image of TSM

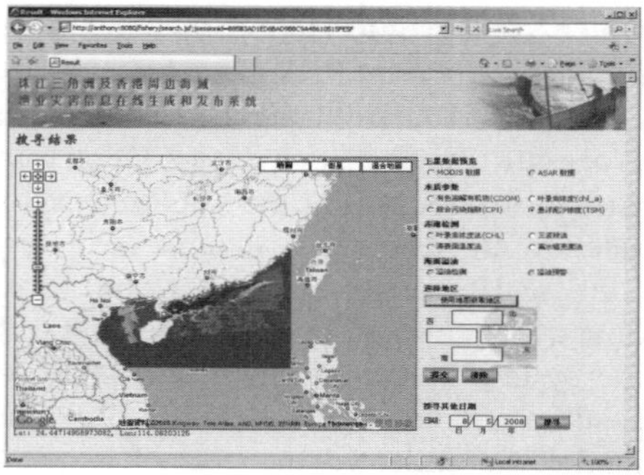

Fig. 3. Browser-end display of TSM spatial distribution generated according to user's requirement

Figure 3 shows the performance displayed in the user-side browser after synthesizing the generated TSM image and Google Map. The TSM image returned from the server in the system uses 'png' format, in which non-marine parts are fully transparent so as not to block the land part on the image. If satisfied with the disaster information displayed in the browser, the registered user can request the generation of a disaster information report such as that shown in Figure 4 which contains auxiliary information, including satellite name, satellite location, imaging time, imaging mode, disaster level (if any), disaster area (if any), disaster trend (if any), and other information.

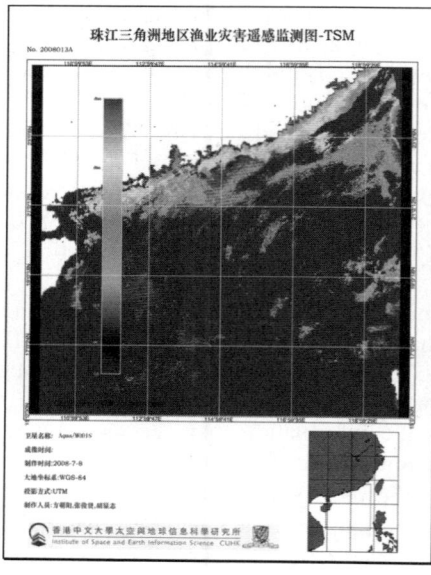

Fig. 4. Detailed disaster report with auxiliary information

5 Conclusion

In this paper, a system framework – design and application of DMWIS is introduced for online generation and distribution of disaster information based on satellite remote sensing data. In this way, the objective of near-real-time monitoring and early warning of natural disasters is achieved. The fisheries disaster monitoring and early warning information system for Pearl River Delta region was developed. The successful development of the system indicates the effectiveness of framework supported near-real-time online generation and distribution of disaster information.

The design and realization of satellite remote sensing data based online generation and distribution system can promote and accelerate the operation of the researches of satellite remote sensing. The system makes ordinary users able to customize the needed disaster information via WEB network. It has the following four characteristics: First, users can achieve personalized customization of near-real-time online disaster information. Second, the image with a size of dozens of KB can describe in detail the needed thematic disaster information. Compared with the satellite remote sensing data of hundreds of MB, the small size of image can be transmitted more rapidly in the WEB environment. Third, users only get the required thematic disaster information in custom-side browser while the raw data are isolated in the server side by the firewall which guarantees the security of data. Lastly, users get the disaster information remotely via the Internet so no substantial investments will be needed for hardware, software or personnel.

Acknowledgement

The work described in this paper is jointly funded by Hong Kong Innovation and Technology Fund (ITF, GHP/026/06), and Social Science and Education Panel Direct Grant of Chinese university of Hong Kong (2020903). The authors wish to take this opportunity to express their sincere acknowledgment to them.

References

1. Alonso, G., Abbadi, A.: Cooperative Modeling in Applied Geographic Research. International Journal of Intelligent and Cooperative Information Systems 3(1), 83–102 (1994)
2. Alonso, G., Hagen, C.: Geo - Opera: Workflow Concepts for Spatial Processes. In: The Proceedings of Advances in Spatial Databases (1995)
3. Development and Reform Commission of Guangdong Province, The 11th Five-Year Development Plan of Guangdong Province economic and social development (2006)
4. Feng, S.Z., LI, F.Q., Li, J.: An Introduction to Marine Science. High Education Press, Beijing (2000)
5. Fang, C.Y., Chen, G.: A TOPEX-Based Marine Geographic Information System for Prediction of Extreme Sea Surface Wind Speeds. Journal of Ocean University of Qingdao, 789–794 (2002)
6. Fang, C.Y., Lin, H., Eric, G., Chen, G.: A Satellite Remote Sensing Based Marine and Atmospheric Spatio-temporal Data Model, In: Proc. of SPIE, Geoinformatics 2006: Geospatial Information Science, vol. 6420 (2006)

7. Fang, C.Y., Lin, H., Tang, D., Pang, Y.: Development of the Disaster Monitoring and Warning Information System for the Fishery Industry in the Pearl River Delta Region. In: Proceeding of ACRS2007, Kualalampur, Malaysia (2007)

8. Chen, G., Fang, C.: Application of RS and GIS Technologies in the Analysis of Global Sea Surface Wind Speed. Journal of Remote Sensing 6(2), 123–128 (2002)

9. Green, D.R.: Cartography and the Internet. The Cartographical Journal 34, 23–27 (1997)

10. https://archive.iseis.cuhk.edu.hk/envisat/eng/ (2008)

11. http://modis.gsfc.nasa.gov/ (2008)

12. http://modis.gsfc.nasa.gov/gallery/# (2008)

13. http://oceancolor.gsfc.nasa.gov/cgi/image_archive.cgi?c=FIRE (2008)

14. http://www.nodc.noaa.gov/SatelliteData/pathfinder4km/ (2008)

15. Huang, B., Lin, H.: A Java/CGI approach to developing a geographic virtual reality toolkit on the Internet. Computers & Geosciences 28, 13–19 (2002)

16. Ji, W., Li, R.: Marine and Coastal GIS: Science or Technology Driven? Marine Geodesy 26, 1–2 (2003)

17. Lockwood, M., Li, R.: Marine geographic information systems: What sets them apart? Marine Geodesy 22, 157–159 (1995)

18. Oštir, K., Veljanovski, T., Podobnikar, T., Stancic, Z.: Application of satellite remote sensing in natural hazard management: the Mount Mangart landslide case study. International Journal of Remote Sensing 24(20), 3983–4002 (2003)

19. Peng, Z.R., Tsou, M.S.: Internet GIS: Distributed Geographic Information Services for the Internet and Wireless Networks. Wiley, New York (2003)

20. Su, Y., Slottow, J., Mozes, A.: Distributing proprietary geographic data on the World Wide Web—UCLA GIS database and map server. Computers & Geosciences 26, 741–749 (2000)

21. Weske, M., Vossen, G., Mederos, B.: Workflow Management in Geoprocessing Applications. In: Proc. 6th ACM International Symposium Geographic Information Systems - ACMGIS 1998, pp. 88–93 (1998)

22. Wright, D.J., O'Dea, E., Cushing, J.B., Cuny, J.E., Toomey, D.R.: Why Web GIS Not Be Enough: A Case Study with the Virtual Research Vessel. Marine Geodesy 26, 73–86 (2003)

A Toponym Resolution Service Following the OGC WPS Standard*

Susana Ladra, Miguel R. Luaces**, Oscar Pedreira, and Diego Seco

Database Laboratory, University of A Coruña
Campus de Elviña, 15071 A Coruña, Spain
{sladra,luaces,opedreira,dseco}@udc.es

Abstract. In the research field of Geographic Information Systems (GIS), a cooperative effort has been undertaken by several international organizations to define standards and specifications for interoperable systems. The Web Processing Service (WPS) is one of the most recent specifications of the Open Geospatial Consortium (OGC). It is designed to standardize the way that GIS calculations are made available to the Internet.

We present in this paper a WPS to perform *Toponym Resolution*. This service defines two geospatial operations. The first operation, *getAll*, returns all possible geographic descriptions with the requested name ordered by a relevance ranking. The second operation, *getMostProbable*, filters the result and returns only the most probable geographic description. Furthermore, both operations can be parameterized according to the level of detail needed in the result.

Keywords: Web Services, Open Geospatial Consortium, Web Processing Service, Toponym Resolution.

1 Introduction

The research field of Geographic Information Systems [1] has received much attention during the last years. Recent improvements in hardware have made the implementation of this type of systems affordable for many organizations. Furthermore, a cooperative effort has been undertaken by two international organizations (ISO [2] and the Open Geospatial Consortium [3]) to define standards and specifications for interoperable systems. This effort is making possible that many public organizations are working on the construction of spatial data infrastructures [4] that will enable them to share their geographic information.

* This work has been partially supported by "Ministerio de Educación y Ciencia" (PGE y FEDER) ref. TIN2006-16071-C03-03, by "Xunta de Galicia" ref. PGIDIT05SIN10502PR and ref. 2006/4, by "Ministerio de Educación y Ciencia" ref. AP-2006-03214 (FPU Program) for Oscar Pedreira, and by "Dirección Xeral de Ordenación e Calidade do Sistema Universitario de Galicia, da Consellería de Educación e Ordenación Universitaria-Xunta de Galicia" for Diego Seco.

** Corresponding author.

M. Bertolotto, C. Ray, and X. Li (Eds.): W2GIS 2008, LNCS 5373, pp. 75–85, 2008.

The OGC is a consensus standards organization that is leading the creation of standards to allow the development of interoperable geospatial systems. One of the most recent specifications standardized by the OGC is the Web Processing Service (WPS) [5] (version 1.0.0 of this standard was published on June, 2007). The WPS specification defines a mechanism by which a client can submit a spatial processing task to a server to be completed. In other words, this specification standardizes the way that GIS calculations are made available in Internet. In this paper, we briefly summarize the most important characteristics of the specification and we present a *Toponym Resolution* service developed according with its interface.

Toponym Resolution is a task related to mapping a place name to a representation of the extensional semantics of the location referred, such as a geographic latitude/longitude footprint [6]. This task has been widely used in *Geographic Information Retrieval* (GIR), *question answering*, or *map generation*. The research field in GIR [7] has appeared a few years ago as the confluence of *Geographic Information Systems* [1] and *Information Retrieval* [8]. The main goal of this field is to define index structures and techniques to efficiently store and retrieve documents using both the text and the geographic references contained within the text. Therefore, the documents have to be annotated with the toponyms mentioned in the text. This task has recently been automated, achieving near-human performance using machine learning [9]. However, annotating the documents with the toponyms mentioned in the text is not enough when the documents have to be spatially indexed. In this case, place names must additionally be related to a correlate in a model of the world (for example, using its coordinates in latitude/longitude). A *gazetteer* could be used to obtain these *geo-references*.

A gazetteer is a geographic dictionary that contains, in addition to location names, alternative names, populations, location of places, and other information related to the location. However, Gazetteers are not enough to fully automate the geo-referencing task because they provide the toponyms and the coordinates associated with a place name without any measure of relevance. This problem is related with the *referential ambiguity*. For example, *London* is the capital of the United Kingdom but it is a city in Ontario, Canadaa too. Given the question *where is London*, a Gazetteer would return both locations without giving any hint of which of them is more appropriate.

Furthermore, gazetteers do not usually provide geometries for the location names other than a single representative point (its coordinates). But, sometimes, the real geometry of the toponym is needed. In [10], the authors describe a spatial index structure where the nodes of the structure are connected by means of inclusion relationships. Therefore, each non-leaf node stores, as well as the toponym, the bounding box of the geometry. For such an application, the authors need a service that returns not only the most probable location, but also its complete geometry to build the spatial index. In this paper, we present a service to perform *Toponym Resolution*. This service provides an operation to obtain all the possible geographic descriptions for a toponym ordered by a ranking of relevance. Moreover, the service provides an operation to obtain only the most probable geographic description. Both operations can be parameterized according to the level of detail needed in the result (i.e., whether a single representative point is enough or a complete geometry is needed). In accordance with the current trend in GIS, these operations, or spatial processes, are offered as a service according with the WPS specification. The rest of the paper is organized as follows. We first describe some related work in Section 2. In addition to that, Section 3 resumes the main characteristics of the WPS specification. Then, in Section 4, we present the general architecture of the system and describe its components. After that, in Section 5, some implementation details are described. Finally, Section 6 presents some conclusions and future lines of work.

2   Related Work

Gazetteers are considered one of the most important components in Spatial Data Infrastructures [4]. A gazetteer service returns information about places in response to queries using their identifiers (e.g., location names). This information typically contains geographic data, such as the coordinates, social statistics, etc. The international OGC specification *Gazetteer Profile of WFS* (WFS-G) [11] standardizes the functionalities that may implement a gazetteer. Service metadata, operations, and types of geographic entities are defined in this specification. The main differences between the WFS-G and WFS specifications are:

- The gazetteer structure is described in an additional section of the document describing service metadata.
- All the geographic entities defined in a WFS-G are subclass of the predefined *SL_LocationInstance*. Therefore, geographic entities share a set of basic attributes and can define other attributes specifically designed for the concrete application.

Many free resources have been published in Internet that provide gazetteer functionalities, geographic ontologies, etc. *Alexandria Digital Library* [12], *Getty Thesaurus of Geographic Names* [13], or *GeoNames* [14] are some examples of these resources. However, none of them define a service following the WFS-G specification.

An important drawback of the gazetteers is that they do not usually provide a complete geographic description of the location returned by a query. There are several cartographic resources that can be used to complete the information provided by the gazetteers. Global Administrative Unit Layers (GAUL) [15] and *Vector Map* (VMAP) cartography [16] are two interesting resources because they provide a complete and updated cartography of the world. However, this cartography is not usually offered by gazetteers. Instead, only a single representative point is returned for each location queried.

Gazetteers are a key component in the task of *Toponym Resolution*. The goal of this task is to obtain the *referent* of the place names. The work of Leidner

[6] in this task is focused on the research field in Geographic Information Retrieval (GIR). Several papers describe the architecture of GIR systems and the NERC+R (Named Entity Recognition and Classification with Resolution) task is shared in most of the proposals. Recognizing the toponyms in the texts of the documents and relate these toponyms to correlate in a model of the world is the main goal of this task. Some papers that deal with different aspects of this problem in the context of GIR have been published in the last years [17] [18] [19]. Web-a-where [17] uses *spatial containers* in order to identify locations in documents, MetaCarta (the commercial system described in [18]) uses a natural language processing method, and STEWARD [19] uses an hybrid approach. A common drawback of gazetteers when applied to this task is that, given a location name, gazetteers provide a list of toponyms that is not ordered by relevance. Therefore, the user of the gazetteer must find a method to order the list of results.

3   OGC Web Processing Service

The Web Processing Service (WPS) [5] is one of the most recent specifications of the OGC. This standard defines a mechanism by which a client can submit a spatial processing task to a server to be completed. Recently, some papers have appeared that review the specification and propose several examples of its usage [20] [21]. In this section we briefly resume the most important characteristics of the specification.

As said above, the WPS specification is centred in the communication between the server and the client. An XML-based protocol using the POST method of HTTP has been defined to perform this communication. Furthermore, requests can also be expressed in Key-Value-Pairs (KVP) encoding using the GET method of HTTP. In addition to the communication protocol, the specification defines three operations:

- *GetCapabilities*. This operation is common in many OGC specifications. The response is a XML document with two main parts: *ServiceIdentification* and *ProcessOfferings*. The first one is shared with other OGC specifications and it describes information of interest regarding the service provider. The second one lists all the processes offered by the service.
- *DescribeProcess*. After a client parses the GetCapabilities response, it has a list of the processes offered by the service. The operation *DescribeProcess* can then be used to request more information about each of them. This operation receives the process identifier as a parameter and returns a XML document that describes all the characteristics of the process such as the title, abstract, etc. Moreover, a full description of the process input parameters is provided in order to allow the client to understand the way in which the process is invoked. Finally, the response document also describes the output format of the process.
- *Execute*. Finally, clients have enough information to request for the execution of a process. The response of the operation *Execute* is a XML document with

information about the *status* of the process, inputs that were used, and the output. The output can be a simple literal (for example, a numerical result or the url where a complex document is accessible) or a complex output (for example, a feature collection description in GML [22]).

Geospatial processes can be very complex and they can take a long time to complete (in terms of hours, days, or even weeks). Therefore, these processes must be performed in an asynchronous way. The specification defines the *status* description in the XML document in response to a *Execute* request for this purpose. The value *ProcessAccepted* indicates that the process was correctly received. *ProcessStarted* indicates that the server is performing the process. *ProcessSucceeded* indicates that the process is finished, and therefore, the result is ready. Finally, *ProcessFiled* indicates that a problem appeared in the execution of the process.

One of the most attractive characteristics of this specification is that it can be applied to an unlimited number of cases. All geospatial process can be offered in Internet following this specification. However, there are some issues that must be considered to decide whether to define a process as a WPS or not. First, complex processes that take a long time to complete are the best candidates to be implemented as a WPS. However, if the complexity of the process is low and the main part of the time is taken up by managing a lot of data stored locally and not on the server, the process can be completed more effectively locally. Second, WPS have the advantages of all the general-purpose web services. One of the most important is that the service is centralized. Therefore, a WPS is appropriate for the deployment of new processes that are under active development. WPS developers can release new versions simply updating the version of the process implementation in the server. Finally, WPS makes possible to create advanced services by means of the *orchestration* of several services.

Recently, several frameworks and implementations of the OGC WPS have appeared to make it usage easier. However, most of them are implementing the 0.4.0 version of the standard. We use the *52 North WPS* [23] framework in the implementation of our system. The 52 North Web Processing Service enables the deployment of geo-processes on the web. It features a pluggable architecture for processes and data encodings. The implementation is based on the version 1.0.0 of the specification.

4   System Architecture

Fig. 1 shows our proposal for the system architecture of a *Web Processing Service* (WPS) to perform *Toponym Resolotuion*. The architecture has two independent layers: the *WPS layer* and the *Toponym Resolution layer*.

The WPS layer is at the top of the architecture. As we noted before, the 52 North WPS implementation [23] has been used in this work. In [24], the authors present the 52 North WPS architecture and an example of use for a generalization process. This architecture is quite simple. There is a *Request Processor*

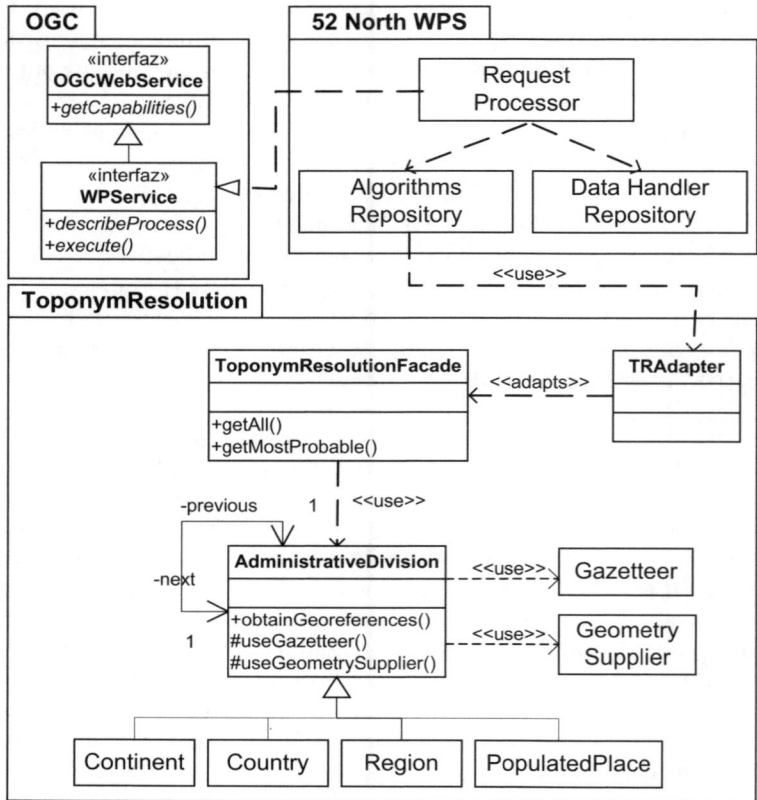

Fig. 1. System Architecture

that manages the communication protocol with the clients. This *Request Processor* implements the OGC WPS specification and it encapsulates all the details related to the communication protocol. In order to achieve a high extensibility, the implementation of a 52 North WPS is organized in *repositories* that provide dynamic access to the embedded functionality of the WPS. For each geospatial process offered by the service, an algorithm must be defined in the *Algorithms Repository*. For example, the algorithms repository in [24] is composed of several generalization algorithms. In our implementation, we adopt the notion of *repository* and we design an intermediary component to adapt the geospatial algorithm to the interface of the specific repository (see the design pattern *Adapter* in [25]). Therefore, our particular implementation of the algorithms does not depend excesively on the details of the *Request Processor*.

The bottom part of the figure shows the architecture of the Toponym Resolution component. The *TRAdapter* class represents the *adapter* between this component and the algorithms repository of 52 North. The adapter uses the *ToponymResolutionFacade* that is a *Facade* [25] that provides a simplified interface of the component. This facade defines two public operations: *getAll* and *get-MostProbable*. The first one returns all the possible geographic descriptions in response to a place name. There are two main differences between this operation and the functionality offered by a gazetteer. First, our implementation can be configured to obtain the real geometry, the *bounding box*, or a single representative point. Second, we provide these descriptions ordered by a relevance ranking. The second operation returns only the most probable geographic description.

The implementation of both operations uses a hierarchy of *Administrative Divisions* to perform the process that defines four levels of administrative divisions (*Continent, Country, Region,* and *PopulatedPlace*). The implementation is easily extensible because several design patterns were used to obtain a robust and extensible architecture. First, the hierarchy follows the pattern *Chain of Responsibility*. Therefore, the class that represents each administrative level is in charge of a part of the whole process and it delegates the rest on the next level. Two *chains of responsibility* are configured in the system. The first one is used to go down the hierarchy finding place names. Once a toponym is found, the second chain is used to go up the hierarchy in order to return the *complete path* that fully describes the toponym in the hierarchy. For example, if the requested place name is *A Coruña*, the complete path is composed by the geographic descriptions of *Europe, Spain, Galicia,* and *A Coruña*. Furthermore, algorithms to obtain the georeferences were designed following the pattern *Template Method*. Therefore, the superclass (*AdministrativeDivision*) defines the general algorithm and the concrete steps may be changed by the subclasess. These steps define how the *Gazetteer* and the *Geometry Supplier* are queried in each level. In the section 5, we present more details about the concrete algorithm implemented in the system to retrieve and rank toponyms.

5   Implementation

As we said in the previous section, the *Toponym Resolution* layer uses a *Gazetteer* and a *Geometry Supplier* in order to obtain the geographic descriptions. In our test implementation we use *Geonames* [14] that provides a geographical database available under a creative commons attribution license. This database contains more than two million populated places over the world with their latitude/longitude coordinates in WGS84 (*World Geodetic System 1984*). All the populated places are categorized so that it is possible to classify them into the different administrative division levels that are defined by the architecture (continents, countries, regions, and populated places).

However, Geonames (and Gazetteers in general) does not provide geometries for the location names other than a single representative point. But for our system we need the real geometry of the location name. We define a *Geometry Supplier* service to obtain the geometries of those location names. As a base for this service we used the *Vector Map* (VMap) cartography [16]. VMap is an updated and improved version of the National Imagery and Mapping Agency's Digital Chart of the World. It supplies first and second level administrative

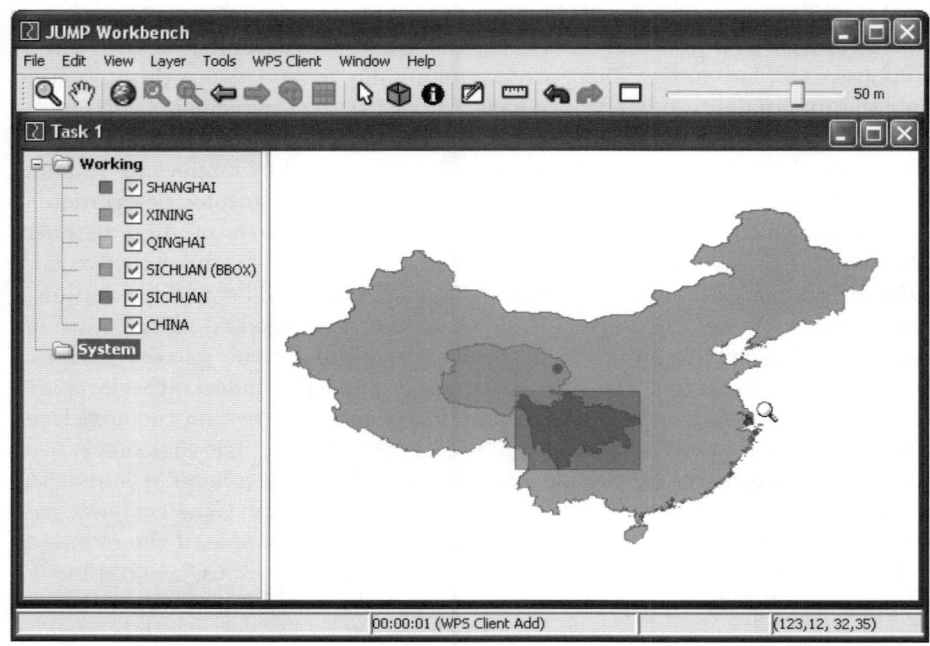

Fig. 2. Examples of results using the WPS plugin for JUMP

division geometries in a proprietary format. However, there are free tools that can create *shapefiles* from that format, such as FWTools [26]. We have created a PostGIS [27] spatial database with these *shapefiles* and we have done several corrections and improvements over this database.

Even though our test implementation uses Geonames and VMap, it has been designed so that these components are easily exchangeable. All accesses to these components are performed through generic interfaces that can be easily implemented for other components.

The spatial operations defined by the hierarchy in the architecture combine both services to geo-reference location names. Each level contains a connection to the gazetteer and to the geometry supplier in order to retrieve the data needed by the process. In other words, subclasses in this hierarchy change the abstract methods of the superclass to implement real queries to both services.

Furthermore, the algorithm to obtain geo-references is implemented in two steps each of them using one of the *chain of responsabilities* defined in the hierarchy. In the first step, each level obtains from the gazetteer all the locations with the requested name. After that, in the second step, the system builds the complete path of geographic descriptions from bottom to top. For instance, if the requested location name was London, in the first step the system obtains at least two locations with this name. After that, it returns the paths *United Kingdom, England, London* and *Canada, Ontario, London*. Finally, to elaborate a relevance ranking of the results (or to return the most relevant result) the algorithm computes a measure of relevance for each result. This measure combines the length of the path, the population of the place, a weighting factor depending of wheter the place is a capital, main city, etc. Most of these data come from the gazetteer.

Fig. 2 presents some examples using the generic WPS plugin for JUMP [28] developed by 52 North. JUMP provides a graphic user interface for viewing and manipulating spatial data-sets. The architecture of this tool is very extensible and it defines a mechanism of extension based on plugins. 52 North developed a plugin for JUMP that implements a generic WPS client. We use this client to test the Toponym Resolution WPS.

One can see in the figure the result of several requests to the *getMostProbable* operation. All of these requests were executed with the parameter *full_path* set to *false*. If this parameter is set to *true*, the WPS returns the geographic description of all the nodes in the path (continent, country, etc.). Furthermore, all the requests, except the layer namely *SICHUAN(BBox)*, were executed with the parameter *bounding_box* set to *false*. The bounding box, instead the real geometry, is returned by the WPS if this parameter is set to *true*. The requested place names are, from bottom to top, *China* (a country), *Sichuan* (a province of China), the bounding box of this province, *Qinghai* (a province of China), *Xining* (the capital of Qinghai), and *Shanghai* (the host city of the conference W2GIS 2008).

6   Conclusions and Future Work

We have presented in this paper a system to perform *Toponym Resolution*. The interface of this system defines two spatial operations *getAll* and *getMostProbable*. The first one returns all the geographic descriptions with the requested place name ordered by a relevance ranking. The second one filters the result and it returns only the most relevant geographic description with the requested name. Furthermore, both operations can be customized with two parameters. The *bbox* parameter is used to obtain the bounding box of the geometries instead of the real geometries. The *full_path* parameter is used to obtain the full path that represents the requested place name instead of the leaf node of this path. Moreover, following the current trend in GIS, we developed a Web Processing Service (WPS) to offer both *getAll* and *getMostProbable* operations as processes that can be performed through the Internet.

Future improvements of this WPS are possible. Many times, intrinsic features of the toponyms (such as population or administrative level) are not enough to decide the most relevant result in a certain *context*. For example, if the place name *Santiago* appears in a document with other place names such as *Atacama*, or *Magallanes*, the document describes regions in Chile. However, if *Santiago* appears with *Madrid* or *Barcelona*, the document describes places in Spain. We are currently working on a new operation that can be invoked with more than one place name. The result of this operation must be the most probable geographic descriptions to each place name. This operations can be very useful in the research field of *Geographic Information Retrieval* (GIR). Therefore, another line of future work involves integrating this WPS in the architecture of GIR systems. Furthermore, changes in the algorithms are needed to improve the performance of the system. Finally, we plan on exploring other gazetteers and cartographies to determine the way that they affect the performance of the system.

References

1. Worboys, M.F.: GIS: A Computing Perspective. CRC Press, Boca Raton (2004)
2. ISO/IEC: Geographic Information – Reference Model. International Standard 19101, ISO/IEC (2002)
3. Open GIS Consortium, Inc.: OpenGIS Reference Model. OpenGIS Project Document 03-040, Open GIS Consortium, Inc. (2003)
4. Global Spatial Data Infrastructure Association: Online documentation (retrieved, May 2007), http://www.gsdi.org/
5. Open GIS Consortium, Inc.: OpenGIS Web Processing Service Implementation Specification. OpenGIS Standard 05-007r7, Open GIS Consortium, Inc. (2007)
6. Leidner, J.L.: Toponym Resolution in text: "Which Sheffield is it?". In: Proceedings of the the 27th Annual International ACM SIGIR Conference (SIGIR 2004), Sheffield, UK (abstract, doctoral consortium)(2004)
7. Jones, C.B., Purves, R., Ruas, A., Sanderson, M., Sester, M., van Kreveld, M., Weibel, R.: Spatial information retrieval and geographical ontologies an overview of the SPIRIT project. In: Proceedings of the 25th Annual International ACM SIGIR Conference on Research and Development in Information Retrieval, pp. 387–388 (2002)
8. Baeza-Yates, R., Ribeiro-Neto, B.: Modern Information Retrieval. Addison-Wesley, Reading (1999)
9. Zheng, G., Su, J.: Named entity tagging using an hmm-based chunk tagger. In: Proceedings of the 40th Annual Meeting of the Association for Computational Linguistics, pp. 209–219 (2002)
10. Luaces, M.R., Paramá, J.R., Pedreira, O., Seco, D.: An ontology-based index to retrieve documents with geographic information. In: Ludäscher, B., Mamoulis, N. (eds.) SSDBM 2008. LNCS, vol. 5069, pp. 384–400. Springer, Heidelberg (2008)
11. Open GIS Consortium, Inc.: Gazetteer Profile of WFS (WFS-G) Specification. Opengis project document, Open GIS Consortium, Inc. (2006)
12. Library, A.D.: Gazetteer (retrieved September 2007), http://www.alexandria.ucsb.edu/gazetteer/
13. Getty, T.: Getty Thesaurus of Geographic Names (retrieved, September 2007), http://www.getty.edu/research/conducting_research/vocabularies/tgn/
14. Geonames: Gazetteer (retrieved, September 2007), http://www.geonames.org
15. Food and Agriculture Organization of the United Nations (FAO): Global Administrative Unit Layers (GAUL) (retrieved, September 2007), http://www.fao.org/geonetwork/srv/en/metadata.show?id=12691
16. National Imagery and Mapping Agency (NIMA): Vector Map Level 0 (retrieved, September 2007), http://www.mapability.com
17. Amitay, E., Har'El, N., Sivan, R., Soffer, A.: Web-a-where: geotagging web content. In: Proceedings of 27th annual international ACM SIGIR, pp. 273–280 (2004)
18. Rauch, E., Bukatin, M., Baker, K.: A confidence-based framework for disambiguating geographic terms. In: Proceedings of the HLT-NAACL 2003 workshop on Analysis of geographic references, pp. 50–54 (2003)

19. Lieberman, M.D., Samet, H., Sankaranarayanan, J., Sperling, J.: STEEWARD: Architecture of a Spatio-Textual Search Engine. In: Proceedings of the 15th ACM Int. Symp. on Advances in Geographic Infomation Systems (ACMGIS 2007), pp. 186–193 (2007)
20. Michaelis, C.D., Ames, D.P.: Evaluation and implementation of the ogc web processing service for use in client-side gis. Geoinformatica (2008)
21. Cepický, J.: Ogc web processing service and it's usage. In: Proceedings of the 15th International Symposium GIS Ostrava (2008)
22. Open GIS Consortium, Inc.: OpenGIS Geographic Markup Language (GML) Encoding Standard. Opengis standard, Open GIS Consortium, Inc. (2007)
23. 52 North: Geoprocessing (retrieved, December 2007), http://52north.org/
24. Foerster, T., Stoter, J.: Establishing an ogc web processing service for generalization process. In: Proceedings of the Workshop of the ICA Commission on Map Generalisation and Multiple Representation (2006)
25. Gamma, E., Helm, R., Johnson, R., Vlissides, J.: Design Patterns: Elements of Reusable Object-oriented Software. Addison-Wesley, Reading (1996)
26. FWTools: Open Source GIS Binary Kit for Windows and Linux (retrieved, September 2007), http://fwtools.maptools.org
27. Refractions Research: PostGIS (retrieved, June 2007), http://postgis.refractions.net
28. The JUMP Project: JUMP Unified Mapping Platform (retrieved, January 2008), http://www.jump-project.org/

Sensor Web Oriented Web-Based GIS

Shaoqing Shen, Xiao Cheng, and Peng Gong

State Key Lab of Remote Sensing Science, Jointly Sponsored by Institute of Remote
Sensing Applications, Chinese Academy of Sciences, and Beijing Normal University,
3 DaTun Road, Chao Yang District, Beijing, China
shenshaoqing1985@163.com, xcheng@irsa.ac.cn, gong@irsa.ac.cn

Abstract. Web-Based GIS has brought a lot of convenience to the
public. However, traditional Web-Based GIS cannot meet the needs in
applications that request for many different data types and real-time
updates. A Sensor Web is a computer accessible network of many, spatially distributed devices using sensors to monitor conditions at different
locations. In this paper we demonstrate the concept of Sensor Web Oriented Web-Based GIS, which is a new type of Web-Based GIS using
Sensor Web as data source.

We first describe the concept of Sensor Web Oriented Web-Based
GIS after analyzing the disadvantages of traditional Web-Based GIS and
Sensor Web. Then we demonstrate the distributed architecture for such
a system. At last, a prototype application of Sensor Web Oriented Web-
Based GIS is described.

Keywords: Web-Based GIS, Sensor Web, Web Services.

1   Introduction

1.1   Web-Based GIS

Web-Based GIS is GIS combined with internet. Geospatial information can
be published, searched, analyzed, processed, and displayed on the internet on
Web-Based GIS websites [1,2,3,4,5]. This technology supports the sharing of
geospatial information over the Web. However, there are some limitations to
traditional Web-Based GIS. First, spatial database of traditional Web-Based
GIS is static, so traditional Web-Based GIS cannot fulfill the needs of those
applications which request for updating data in real-time. Spatial database of
traditional Web-Based GIS is updated manually. Obviously, for things changing very fast, manual updating is inadequate. Static spatial database cannot
be used to provide support for decision-makers when they are facing problems
varying from minute to minute. For example, during a flood event, traditional
Web-Based GIS cannot tell decision-makers the peak of floods in real-time. If
spatial database can be dynamic in real-time to reflect changes, Web-Based GIS
can help many decision-makers to make faster and more reasonable decisions.

Second, traditional Web-Based GIS does not make full use of data collections
available in the world, which often keeps end-users away from useful information

M. Bertolotto, C. Ray, and X. Li (Eds.): W2GIS 2008, LNCS 5373, pp. 86–95, 2008.

when they make decision. In fact, there are many useful data collections existing in different organizations, institutes, companies and other places. However, it is hard to find, access, and manage them in automatic and generic ways. If people want to use them with Web-Based GIS, they have to search them manually first and then convert them manually to conform data models and data formats which Web-Based GIS could identify, before adding them to spatial database.

In order to improve the performance of Web-Based GIS, it is necessary to find a way to update the spatial database in real-time and also supply it with rich data collections in automatic and general ways. Sensor Web can fulfill this demand.

1.2   Sensor Web

By definition, a sensor is a device that provides a usable output in response to a specific physical quantity, property, or condition which is measured [6].

A Sensor Web is a computer accessible network of many, spatially distributed devices using sensors to monitor conditions at different locations, such as temperature, sound, vibration, pressure, motion or pollutants. A Sensor Web refers to web accessible sensors and archived sensor data that can be discovered and accessed using standard protocols and application program interfaces (APIs) [7,16].

Sensor Web can include both in situ sensors and remote sensors, stationary sensors and mobile sensors, wired sensors and wireless sensors, etc. For example, flood gauges, air pollution monitors, stress gauges on bridges, mobile heart monitors, webcams, satellite-borne earth imaging devices and countless other sensors and sensor systems. In another word, Sensor Web weaves an electronic skin of the Earth, offering full-dimensional, full-scale, and full-phase sensing and monitoring at all levels, global, regional and local [8].

It is predicted that Sensor Web will be one of the top ten technologies to change the world in the twenty first century. Internet has reformed the way people communicate with each other. However, Sensor Web will reform the way people interact with nature because it connects the logic information world to the physical world.

It is natural and necessary for data collected by Sensor Web to have a space property. First, any phenomenon measured by sensors must occur at a particular location. Location is an enabling key to make sensor observations meaningful. As the technology advances in communication and sensor development, we will see more and more sensor systems equipped with GPS that can locate themselves while collecting data [8]. Second, any property of a feature of interest may vary in space. Therefore, when the property is observed, data collection about this property should be a function of space, which is a coverage, such as satellite images. These collections contain space information inside [7].

Therefore, an infrastructure that could store, disseminate, exchange, manage, display, and analyze spatial information is vital for the Sensor Web. Web-Based GIS can be used to achieve this goal.

1.3   Related Work

On the one hand, Sensor Web could provide Web-Based GIS with abundant data collections, both real-time and archived, in an automatic and ubiquitous way. On the other, Web-Based GIS could store and present data collected by Sensor Web with spatial meaning, enable data analysis and supply timely information to decision-makers. It is reasonable to build up a Sensor Web Oriented Web-Based GIS. Sensor Web Oriented Web-Based GIS will be a powerful tool for decision-makers when they are facing complex problems.

Web-Based GIS has been developed for a long time and formed various specifications. Fundamental sciences and technologies supporting Sensor Web, including communication, networking, sensor technology, embedded systems have also matured in the past 15 years. The primary difficulty here is to build a Sensor Web Oriented Web-Based GIS to solve the problems on how to realize the concept of Sensor Web and how to design an architecture to combine Web-Based GIS and Sensor Web.

Relevant research has been done. Some researchers focus on the ubiquitous architecture for any data collection by Sensor Web, while others emphasize the space property of data collection and pay special attention to the space property.

Space property not-concerned architecture. Gibbons and his cooperators designed an agent-based architecture for Sensor Web named 'IrisNet'. This architecture targets on wide-area distributed network of high-bit-rate sensors and is divided into two tiers: SAs (Sensing Agents), where each SA provides, preprocesses and reduces raw data from a physical sensor, and OAs (Organization Agents), where each OA provides a sensor service. SAs have a generic data acquisition interface to access sensors. Wide-area distributed OAs are orchestrated in a hierarchy. An important characteristic of this architecture is that sensor agents are dynamically programmable. A new service, i.e. an OA can send preprocessing steps, called a senslet, to a sensor agent. The sensor agent can then execute these steps on its raw data stream and return the results. The designers realize three prototype applications to test the architecture: parking-space finder, network and host monitor (IrisLog) and coastal imaging service in collaboration with oceanographers of the Argus project at Oregon State University [17].

Biswas constructed an agent-based middleware that bridges the gap between the programmable application layer consisting of software agents and the physical layer consisting of sensor nodes and tested it in an integrated target tracking experiment. There are three layers: the data layer which consists of sensors that act as the data source, the application layer which fuses and interprets the information, and the service layer which acts as an integration layer between the physical data layer and the software application layer [18].

Moodley proposed an architectural framework, the Sensor Web Agent Platform (SWAP). SWAP allows for integration of arbitrary sensors or sensor networks into a loosely coupled higher level environment that facilitates developing and deploying end user applications across multiple application domains. An ontology framework and an abstract hierarchy are proposed for fusing and integrating data and storing and re-using generic information extraction techniques within the system. One important characteristic of SWAP is that it makes use of both Web Services and Multi Agent Systems [9].

Space property concerned architecture. The Sensor Web Enablement (SWE) initiative proposed by the Open Geospatial Consortium (OGC) extends prominent OGC Web services by providing additional services for integrating Web-connected sensors and sensor systems. The goal of SWE is to enable all types of Web and/or Internet-accessible sensors, instruments, and imaging devices to be accessible and, where applicable, controllable via the Web. The vision is to define and approve the standards foundation for "plug-and-play" Web-based sensor networks. SWE currently defines four Web service specifications and two models and encodings for observations and sensors respectively. The Web services, known as Sensor Observation service for data access, Sensor Planning Service for sensor tasking and feasibility studies, Sensor Alert Service for registering atomic conditions and push based notification, and Web Notification Service as a data transport protocol transformer describe the operational part of the framework. The data models and encodings Observation & Measurement (O & M) and Sensor Model Language (SensorML) are used as data and metadata exchange protocols [7].

Liang and his colleagues proposed to build a geospatial information infrastructure for Sensor Web. The infrastructure is divided into three layers: Sensor Layer, Communication Layer and Information Layer. They use a 'Web Services' approach to build a prototype application named GeoSWIFT Sensing Services. Web Services supports Web-based access, easy integration, and service reusability, thus satisfying their requirements to build a geospatial infrastructure for Sensor Webs with openness, interoperability and extensibility. The architecture advocates use of OGC standards for integrating and exposing sensor data [8].

Nolan developed a multi-agent-based infrastructure named 'AIGA' that supports image processing, geospatial data processing as well as text mining. The architecture focuses on performance and scalability issues related to using mobile agents for image processing, geospatial processing and text mining applications, especially the agent mobility and scalable discovery aspects. The architecture differentiates between data agents and processing agents and coordination agent that provide directory services. Agents communicate with each other through a shared communication space. The system provides a user interface for end-users to compose workflows of agents to provide different application functionality. These workflows can be saved in the system and re-used at a later time [19].

2   The Distributed Architecture of Sensor Web Oriented Web-Based GIS

From the previous work above, we can see that there are two promising technologies for building distributed systems: Web Services and Agent Systems. Because OGC, as the world leading geospatial industry standards organization, uses 'Web Services', we chose 'Web Services' to build Sensor Web Oriented Web-Based GIS.

2.1 Web Services

Web Services describe a standardized way of integrating Web-Based applications using the XML, SOAP, WSDL and UDDI open standards over an Internet protocol backbone. XML is used to tag the data, SOAP is used to transfer the data, WSDL is used for describing the services available and UDDI is used for discovering what services are available. Therefore, Web Services satisfy our requirements to build a distributed Sensor Web Oriented Web-Based GIS with interoperability, scalability and serves a perfect foundation to build such a system [15].

There are three roles in Web Services: Web service providers, Web service requesters and Web service registries. Web service providers develop and define Web Services and then publish them to Web service registries or make them available to Web service requesters directly. Web service requesters perform an operation to locate desired services made available by Web service providers and then request those services from either Web service registries or the publishers directly. Web service registries represent searchable directories of service descriptions defined and published by Web service providers.

There are already four open standards about Web Services related to Sensor Web which are developed by OGC: Sensor Observations Service (SOS), Sensor Planning Service (SPS), Sensor Alert Service (SAS), and Web Notification Service (WNS). They are listed in the Table 1. Sensor service providers can publish them to sensor service registries or make them available to sensor service requesters directly.

Table 1. OGC standard Sensor Services [7]

Service	Functions
SOS	Standard web service interface for requesting, filtering, and retrieving observations and sensor system information. This is the intermediary between a client and an observation repository or near real time sensor channel.
SPS	Standard web service interface for requesting user driven acquisitions and observations. This is the intermediary between a client and a sensor collection management environment.
SAS	Standard web service interface for publishing and subscribing to alerts from sensors.
WNS	Standard web service interface for asynchronous delivery of messages or alerts from SAS and SPS web services and other elements of service workflows.

2.2 Distributed Architecture

The distributed architecture for Sensor Web Oriented Web-Based GIS is based on Web Services. It includes three layers: Sensor Web layer, Web-Based GIS layer, and client layer.

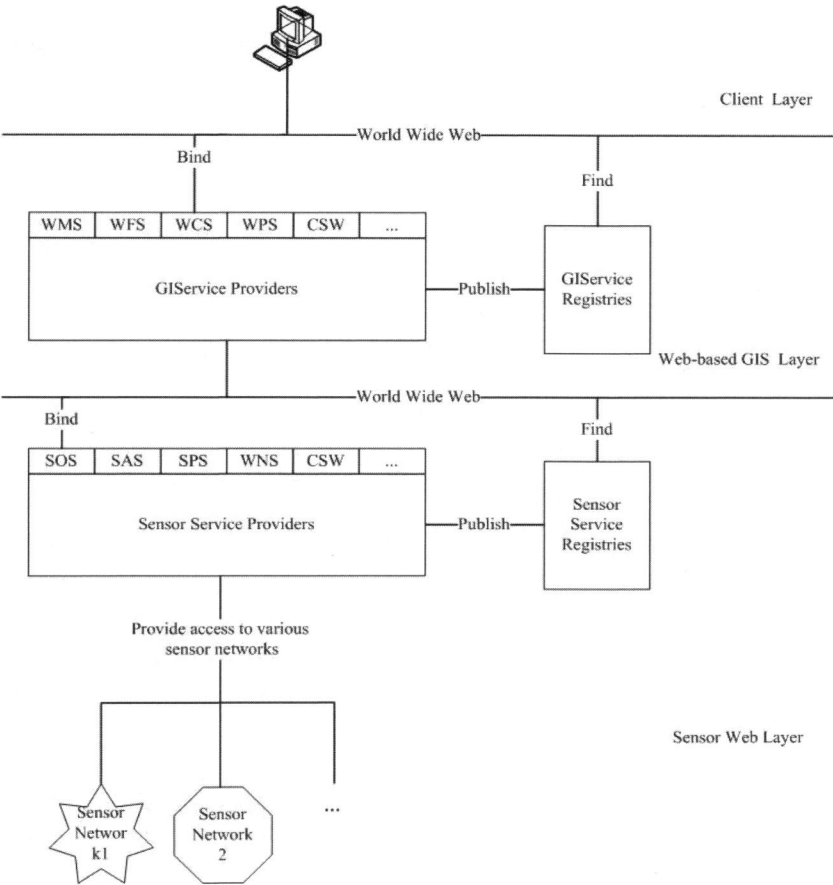

Fig. 1. A general architecture of Web-Based GIS oriented to Sensor Web

Sensor Web Layer. Sensor Web layer is an integrated platform for Sensor Web management. It includes sensor service providers and sensor service registries. Clients on the internet can make use of this layer to locate, access, and control many different sensor networks through open standard interfaces without coping with different communication protocols, different data formats and so on.

Sensor service providers interact with Sensor Web. They provide open standard Web Services related to Sensor Web, such as SOS, SAS, SPS, WNS, and CSW (Catalog Service for Web). They function as wrappers which hide the complexity of various Sensor Webs, such as different communication protocols, different data formats and standards, and provide standard interfaces for clients to collect and access sensor observations and manipulate them in different ways [3]. In this way, various clients only need to follow the standard interfaces provided by the sensor service providers, and are not required to directly deal with the annoying complex Sensor Web. In addition, sensor service providers also publish metadata information and standard interfaces about Sensor Web to sensor service registries. The metadata information includes Sensor Web platform description, Sensor Web description, sensor description, and observation type description and so on.

Sensor service registries provide a mechanism to register and categorize sensor services that are published by sensor service providers and to locate Web Services that users would like to consume. They provide a common mechanism to classify, register, describe, search, maintain, and access information about Sensor Web. So people are easy to locate, access, and use arbitrary Sensor Web through sensor service registries.

Web-Based GIS Layer. Web-Based GIS layer mainly provides GIServices. There are two kinds of GIServices: GIS data Services and GIS function Services. GIS data Services provide spatial data. And GIS functions Services provide operations and processes on spatial data.

Much data used in Sensor Web Oriented Web-Based GIS comes from Sensor Web. Web-Based GIS layer can be sensor services requesters when it requesting data collections from Sensor Web and acquire Sensor Web data collections through standard open sensor services.

Client Layer. Clients on client layer, functioning as Web services requesters, can be as complex as Web Services, or as simple as Browsers. Client layer can manage presentation, interaction, user interfaces, and interactive map manipulation and so on.

3   An Application for Real-Time Water Information System

We developed a system that connects water depth sensors to a Web-Based GIS through wireless sensor network. This Advanced Water Information System (AWIS) aims to monitor in real-time water depths of damaged reservoirs and congested lakes after the Sichuan Earthquake after May 12, 2008. A prototype implementation of the AWIS is currently being developed.

Figure 2 illustrates the prototype application for a real-time water information system. We designed two ways to transmit water depth information. The first is 'base station' to 'cell-phone' through 'GSM' (Global System for Mobile Communication) with the use of GSM module. The second is more complicated: 'base station' to 'server' and then to 'client pc' through 'GSM' and 'Internet' with the use of GPRS module. Users can make use of both ways at the same time or either one. 'Base station' resides on the Sensor Web Layer. It is equipped with GSM and GRPS module for communication. Several sensors in the base station can be mounted to measure different aspects of water bodies as the sensors include pressure sensors, temperature and humidity sensors. Water depth information is obtained by comparing the water pressure with the atmospheric pressure. Other sensors, such as chemical and biological sensors, will be used

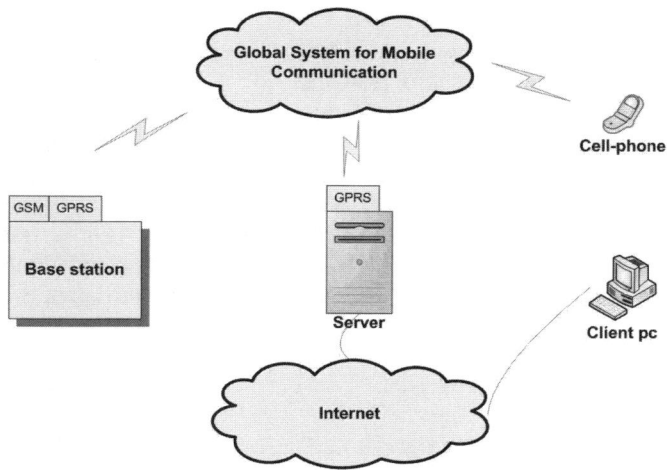

Fig. 2. Hardware structure for a real-time water information system

Fig. 3. Web interface for real-time water monitoring

in the future to evaluate water quality. At present, SOS has been realized to provide observations collected by different sensors to users. Other services will be developed in the future.

The Web-Based GIS Layer is realized in the 'server'. It uses GPRS module to request for observations about water bodies from a Sensor Web layer through standard sensor web services and provides GIServices to 'client pc'. WMS is used to display a map of sensor station deployment. Other GIServices are being developed to integrate models on water conditions with archived and real-time observations to support decision making.

On the Client Layer, 'cell-phone' users will get short messages containing water information. There are two choices for 'cell-phone' users: short messages at certain time interval and short messages by request. 'Pc clients' will have more interactive chances. Figure 3 is the web interface for water depth monitoring.

4   Conclusions

We proposed a conceptual framework of Sensor Web Oriented Web-based GIS to enhance the performance of traditional static Web-based GIS. Web Services are employed to build up a Sensor Web Oriented Web-based GIS because Web Services provide a good dynamic architecture for distributed system construction with interoperability, and scalability. The distributed architecture is divided into three layers: Sensor Web layer, Web-based GIS layer, and client layer. We introduced open Web Services on each layer. The three Layers are integrated into a Sensor Web Oriented Web-based GIS. At last, we describe a prototype application of real-time water information system.

Sensor Web can be employed in an extensive monitoring and sensing system that provides timely, comprehensive, continuous and multi-mode observations. Sensor Web Oriented Web-based GIS will reveal its power when more and more sensors and Sensor Webs are included in such systems.

Acknowledgement

This research is partially funded by an 863 National High Tech Grant (2006AA12Z112) and Scientific Support Grant (2006BAJ01B02).

References

1. Kraak, M.J., Brown, A.: Web Cartography: Developments and Prospects, p. 213. Taylor & Francis, London (2001)
2. Plewe, P.: GIS Online: Information Retrieval, Mapping, and the Internet, Santa Fe, New Mexico, p. 311. OnWord Press (1997)
3. Richard, D.: Development of an Internet atlas of Switzerland. Computers & Geosciences 26(1), 45–50 (2000)
4. Tang, W., Selwood, J.: Connecting Our World: GIS Web Services, Redlands, California, p. 164. ESRI Press (2003)
5. Chen, J., Gong, P.: Practical Geographical Information System. Science Press, Beijing (1999)

6. National Research Council: Expanding the Vision of Sensor Materials p. 146. National Academy Press, Washington DC (1995)
7. Open Geospatial Consortium, Inc., http://www.opengeospatial.org
8. Liang, S.H.L., Croitoru, A., Tao, V.: A distributed geospatial infrastructure for Sensor Web. J.Computers & Geosciences 31, 221–231 (2005)
9. Moodley, D., Simonis, I.: A New Architecture for the Sensor Web: The SWAP Framework. In: 5th International semantic web conference ISWC 2006, Athens, Georgia (2006)
10. Chang, Y.S., Park, H.D.: XML Web Service-based development model for Internet GIS applications. J. International Journal of Geographical Information Science 20(4), 371–399 (2006)
11. Institute of Electrical and Electronics Engineers: IEEE Standard Computer Dictionary: A Compilation of IEEE Standard Computer Glossaries. New York, NY (1990)
12. André, B.B.: Characteristics of scalability and their impact on performance. In: The 2nd international workshop on Software and performance, Ottawa, Ontario, pp. 195–203 (2000) ISBN 1-58113-195-X
13. Kim, D.-H., Kim, M.-S.: Web GIS service component based on open environment. In: Geoscience and Remote Sensing Symposium, IGARSS. IEEE International, vol. 6, pp. 3346–3348. IEEE Press, New York (2002)
14. Lu, X.: An Investigation on Service-Oriented Architecture for Constructing Distributed Web GIS Application. In: IEEE International Conference on Services Computing, SCC. IEEE International 2005, vol. 1, pp. 191–197 (2005)
15. Gardner, T.: An Introduction to Web Services. Ariadne Issue 29 (2001)
16. Gong, P.: Wireless sensor network as a new ground remote sensing technology for environmental monitoring. J. Remote Sensing 11(4), 545–551 (2007)
17. Gibbons, P.B., Karp, B., Ke, Y., Nath, S., Seshan, S.: IrisNet: An Architecture for a Worldwide Sensor Web. J. Pervasive Computing 2(4), 22–33 (2003)
18. Biswas, P.K., Phoha, S.A.: Middleware-driven Architecture for Information Dissemination in Distributed Sensor Networks. In: Intelligent Sensors, Sensor Networks and Information Processing Conference, pp. 605–610 (2004)
19. Nolan, J.J., Sood, A.K., Simon, R.: An Agent-based Architecture for Distributed Imagery & Geospatial Computing. In: Applied Imagery Pattern Recognition Workshop, 2000, pp. 252–258 (2000)

A Design Process for the Development of an Interactive and Adaptive GIS

Mathieu Petit[1], Christophe Claramunt[1], Cyril Ray[1], and Gaëlle Calvary[2]

[1] Naval Academy Research Institute, 29200 Brest Naval BP600, France
{mathieu.petit,christophe.claramunt,cyril.ray}@ecole-navale.fr
[2] Laboratoire d'Informatique de Grenoble, Université Joseph Fourrier, 38041
Grenoble cedex 9 BP53, France
gaelle.calvary@imag.fr

Abstract. Geographical Information Systems (GIS) have long bridged
the gap between geo-information databases and applications. Although
conceptual modelling approaches for GIS have been particularly successful in the representation of the specific properties of geographical
information, there is still a need for a better integration of user intentions and usage. This paper introduces a conceptual framework applied
to GIS and defined not only as a *"Geographic Information System"*, but
also as a *"Geographic Interactive System"*. This approach extends the
conceptual framework of a general purpose mobile interactive system to
the geographical context. Beside a description of user tasks and the domain data layout, the proposed framework considers the geographical
environment as an additional component to the design approach. The
role of the spatial dimension in the design of such an interactive system
is illustrated all along the conception of a real-time ship tracking system.

1 Introduction

Conceptual frameworks applied to the design of geographic information systems are oriented to the modelling and manipulation of spatial data. Conceptual frameworks such as MADS [14] or Perceptory [4] encompass conceptual
modelling approaches that help represent the semantics of geographic data. The
computing architecture implementation, usage context, and users characteristics
restrict to the final steps of the design, or rely on existing client-side applications.
In order to remain generic, domain concepts and their layouts are separated from
the functional and architectural designs of the client side. This level is considered
aside, as a front-end to access and manipulate data [10].

Coming along with the growth of geographical data available through the
Internet, the range of GIS benefits and usages dramatically increases [11]: users
may not be expert of the domain anymore ; platforms that interact with the
system are rich and versatile ; usage contexts differ from one user to another and
GISs might be accessed in mobile contexts, through the mediation of computing
services.

Those emerging constraints make it necessary to adapt current conceptual
GIS frameworks. This leads to a change of paradigm, where a GIS should evolve

M. Bertolotto, C. Ray, and X. Li (Eds.): W2GIS 2008, LNCS 5373, pp. 96–106, 2008.
© Springer-Verlag Berlin Heidelberg 2008

from an *"information system"* to an *"interactive system"*. The design process is still oriented to geographic concepts, but another aspect to consider is the use that can be made of the geographical dimension to enrich user experience and interaction with the system.

A user-oriented and responsive GIS, considered as an interactive system [18], should take into account user goals and tasks. In a related work, a GIS has been considered as an interaction tool between multiple users, and that acts as a support to human dialog [16]. Implicit collaboration between several users favors recommendations of a spatial content for tourism applications [6], or allows cartographic layer selection applied to archaeology [12]. Automated adaptation to hardware resources has also been addressed for cartographic display [9], and for mobile navigation systems [3]. These systems consider the spatial dimension as the core of a user interactive process. They might be considered as *Geographic Interactive systems*, as they take into account users needs regarding their respective environments. Thereby, the design of an interactive GIS is a special case of an interactive system design. It appears that these spatial concepts impact the system conception at the functional and architectural levels.

This paper introduces a framework for the development of a contextual interactive system. It is experimented in section 2 to design a sailing race documentation system. Geographic concepts are part of the framework and provide an additional input to the design process (Section 3). This environment is based on several regions of significance at execution time, whose spatial relationships are likely to generate alternative system behaviors (Section 3.1). The contextual setup of the system refines user tasks and goals (Section 3.2). Section 4 illustrates the benefits of the proposed geographical approach. Finally, section 5 summarizes the contribution and discusses future work.

2   Interactive System Design Principles

At the initial stage of a design process, the user objective should be identified and categorized into a set of *tasks* [1]. These tasks rely on the application domain *concepts*, and qualify the system data and resources.

The user objectives and operative tasks are informally described in a so-called *nominal scenario* (Fig. 1(a)). This documentation is textual based, and summarizes the user needs in the light of the knowledge of the domain experts and system designers. Given a nominal scenario, an interactive system design process follows several conceptual stages:

1. identification and organization of tasks and concepts (Fig. 1(b&c)),
2. software distribution design (Fig. 1(d)), processes and data management implementation (Fig. 1(e)),
3. tasks mapping towards the presentation of functionalities, and user interface development (Fig. 1(h, i and j)).

Nowadays, the wide variety of user platforms available, as well as new ubiquitous usages, stress the need for the integration of *usage contexts* into interactive

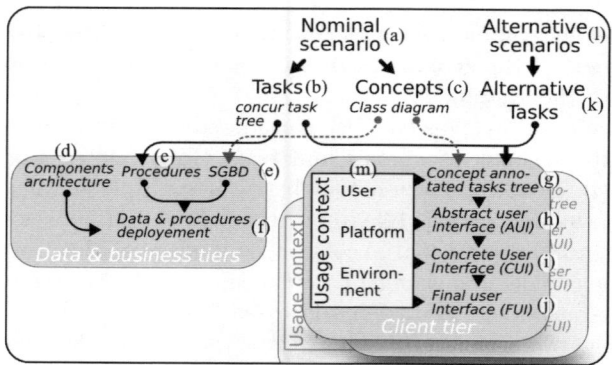

Fig. 1. Design framwork of a context-aware interactive system

system designs [17]. When designing a system, every usage context considered is likely to impact the client-side design (Fig. 1(m)). A usage context is usually described along three orthogonal contexts : user, client platform, and usage environment [8]. Prior to a consideration of this notion of context to a system-wide definition, the next section introduces the early conceptual stage of our case study, from a nominal scenario to the task and concept correlations.

2.1   Nominal Scenario and Sailing Race Documentation System

Our approach is experimented in the context of a sailing championship held once a year in Britany in North-West France. This event allways gathers a large audience, from sail fans to newcomers. The race innings often occurs a long way from the shore, and the audience may only notice and follow closest ships. This entails the need to offer a wireless accessible regatta documentation system to a wider audience.

In the proposed case study, members from the audience and the organizing committee are part of the drafting team. Together, they specify the system using a user-centred design approach and summarize their collaboration within the nominal scenario (Tab. 1). When starting a system design from scratch,

Table 1. Nominal scenario

> "The race documentation system runs on a user's PDA and allows her/him to **follow the _regatta_** in real-time. The PDA provides manipulation tools, and a map of the race area where the racing ships are regularly re-located. The user may **be interested in several ships**, or alternaltively **by other _user interests_**, to set her/his own _area of interest_. If she/he **is interested in a specific ship**, information (_year_, _name_, _crew_ and _pictures_) and real-time data (_location_, _speed_ and _heading_) on this ship are provided. When being close enough to the race area, the user **takes and shares ships _pictures_** with other users."

sequences of tasks and subtasks to meet the user intentions are suggested and integrated into the nominal scenario by designers and HCI experts. In order to to fully comprehend the designed software functionalities, and to complete the task model, the overall design process should be repeated several time. Each iteration adjusts and enriches the prototype, to eventually meet the user needs and intentions [5].

2.2   Domain Tasks and Concepts

A nominal scenario supports the identification of the main concepts and user tasks of the domain. An object modelling formalism lays out the concepts underlined in the nominal scenario. We retained a UML class decomposition as an initial concept structuring template. For example, the concept of "ship" is defined by data like "name", "type" or "year" and is implemented within the class "Boat".

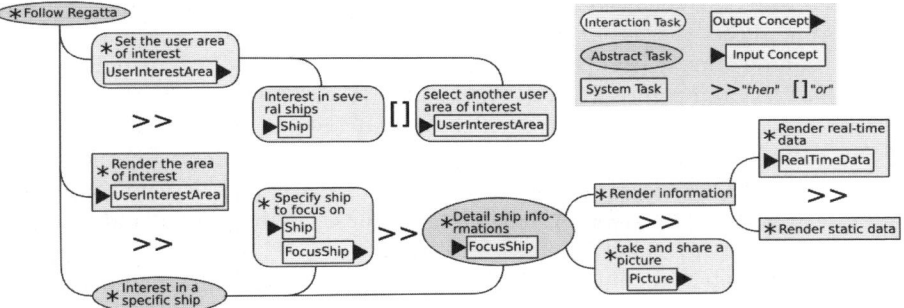

Fig. 2. Task tree of the race documentation system

From the scenario, designers emphasize user tasks. For example, the main task "Follow the regatta" incorporates subtasks such as "Set the area of interest", "Interest in a specific ship" or "take and share pictures". The CTT notation (Concur Task Tree) defines the relationships that organize the tasks, from the user intention level, to the interaction and system levels (Fig. 2)[13]. From a software perspective, each concrete task should be implemented by appropriate procedures and methods, using a semi-automated process [7]. The task tree nodes might be associated with input and output concepts handled by the considered task or sub-task. For example, the interactive task "Set the user area of interest" defines the "UserInterestArea" concept from an input concept, being either a set of ships or another user area of interest.

3   Environment Modelling

To the best of our knowledge, interactive system design methodologies limit the impact of a dynamic context to user-related concepts. The main principle is that a context is worth measurement if it influences the execution of system at the user level. In the described generic framework (Fig. 1), only the final design stages integrate such contextual constraints. Besides, when distributed on a multi-components architecture, an interactive system is related to a changing spatial environment for each of its components.

Therefore, a readily adaptable system strongly relies on accurate environmental conditions, and regularly adapts its functional level at running time. Our extended design framework integrates the system distribution as a foundation for functional behaviour derivation. The *impact of the system spatial environment on the course of the nominal scenario* is grasped at every conceptual stage (Fig. 3).

3.1   Extended Design Framework

In the context of mobility, a runtime environment evolves in an almost continuous mode. These changes are characterized at the design level using a geographical approach to model the system components and their evolution (Fig. 3(a)). This description allows the designer to derive location-based assertions such as "the task T_x can be done within the region R_y" or "the concept C_x is available within the region R_y, and comes from the region R_x"(Fig. 3(b)).

This extended framework derives the set of possible spatial configurations of the environment, according to the spatio-temporal mobility of the system components (Fig. 3(c)). The course of the nominal scenario depends on the validity of these configurations. A description of these environmental states reports the scenario restrictions at the task level (Fig. 3(d)).

Some of the spatial configurations derived do not constraint any part of the nominal scenario. These spatial configurations characterize alternative execution environments. Initially, they were not stated in the nominal scenario, but they might be of interest and integrated into the design process. In order to support these alternative executions, the task tree is provided with new tasks. This implies the presence of inherent *alternatives* to the nominal scenario (Fig.3(f)).

The environment characterization influences the architecture design at the system level, and the interaction design at the client level. This architecture fits the environment when designed to comprehend contextual changes at the components level (Fig. 3(e)). On the other hand, the tasks tree is the primitive of the client side design, where nominal and alternative tasks are integrated within an integrated and consistant user-interface (Fig. 3(g)) [2].

3.2   Runtime Environment

The runtime environment is derived from the spatial distribution of the GIS components. This allows to distinguish several system states, and to update the functionality and data available to the user. In a previous work, several regions of significance have been defined to characterize the system at runtime [15]:

- processing region(s) P_x, where the procedures for the completion of a given task are available to the user,

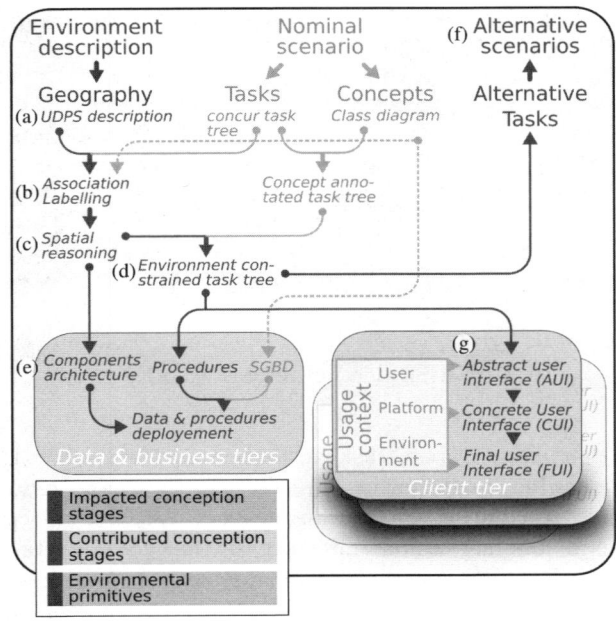

Fig. 3. Extended design framework

- broadcasting region(s) D_x, where the concepts are available to the system,
- user(s) region(s) U_x, where the user is located,
- source region(s) S_x, where the concepts come from.

Significance regions. The set of regions located in space at a given time defines the *runtime environment* of the interactive system. In the case of a sailing race documentation system, data providing heading, speed, and coordinates come from the regatta area $\big($Fig. $4(S_1)\big)$. The region where the user is located is defined by its immediate environment, so called interaction space $\big($Fig. $4(U_1)\big)$.

The processing and broadcasting regions are constrained by the capabilities to physically implement the system components. Real-time positioning data are collected through wireless communications $\big($Fig. $4(\ast)\big)$. During a regatta, sailors are not allowed to access the system concepts. They are broadcasted in a limited region, far from the race area $\big($Fig. $4(D_1)\big)$. The processing procedures on location-based concepts are accessible to the audience, close to the arrival of the regatta $\big($Fig. $4(P_1)\big)$.

Tasks and concept labelling. In order to identify the region in the environment that supports a given task execution, tasks and concepts are *labeled* by their respective regions of influence. Let us consider an interactive task $Task_x$, that processes the concept $Concept_x$ in and out. Assuming that, $Task_x$ and $Concept_x$ are up for usage in regions P_x et D_x, respectively, then, region P_x labels $Task_x$ and region D_x labels $Concept_x$.

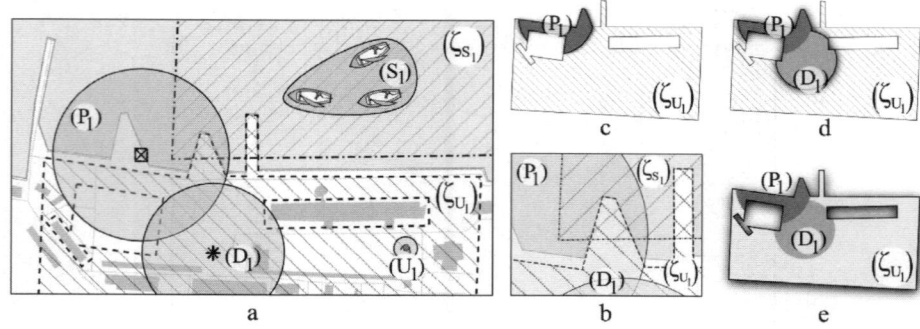

Fig. 4. Significance regions example

Procedures that code a given task perform accurately when their required concepts are accessible. In the proposed example, the task $Task_x$ is *runnable*, provided an access to $Concept_x$. Regarding space, this situation occurs when the tasks and concepts labeling regions intersect. In that case, where $P_x \cap D_x \neq \emptyset$, a user standing in P_x might complete the runnable task $Task_x$.

In the sailing race documentation system, only one processing region and concept broadcasting region are defined. All the task procedures are available in P_1, and every concept of the domain is accessible in D_1. As $P_1 \cap D_1 \neq \emptyset$, the concepts are available inside P_1 whatever the considered task is, and the whole nominal scenario is executable in P_1.

Dynamic environment and spatial reasoning. At the conceptual level, the spatial properties of significance regions characterize the environment variability at execution time. Several alternative system behaviours are derived when the user cannot perform the tasks from the nominal scenario. This gives rise to unhandled environmental configurations, and provide guidelines towards alternative usages recognition [15].

In a changing environment, the regions of significance evolve. Given a region R_x, let the *mobility area* ζ_{R_x} denotes the set of possible locations of R_x during the system uptime. When $R_x \subsetneq \zeta_{R_x}$, the region R_x is *mobile*. Conversely, this region is *fixed* when $\zeta_{R_x} = R_x$. At a given time of the execution, the *system state* is characterized by the set of intersecting regions of significance. For example, the initial system state for the regatta case study $\big($Fig. 4(a)$\big)$ is as follows:

$$\{P_1 \cap D_1 \neq \emptyset, P_1 \cap U_1 = \emptyset, P_1 \cap S_1 = \emptyset, U_1 \cap D_1 = \emptyset, U_1 \cap S_1 = \emptyset, D_1 \cap S_1 = \emptyset\}$$

A tabular notation summarizes this formalism $\big($Fig. 5(a-*first matrix*)$\big)$. Per convention, a black cell represents an intersection between the regions of significance.

Any system described by evolving regions generates a countable set of system states. This gives the boundaries of the whole range of spatial configurations.

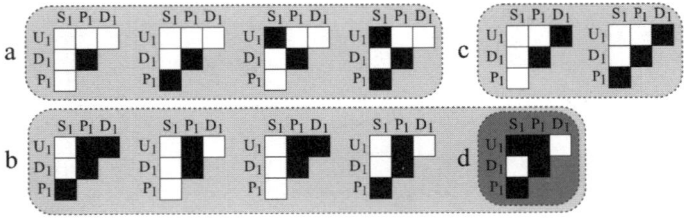

Fig. 5. Spatial configurations of the sailing race system

The system is *highly constrained* when every mobility area ζ_{R_x} with $R_x \subsetneq \zeta_{R_x}$ stands apart, and when the intersecting regions are fixed. In that case, the system is characterized by a spatial configuration and a system state. When all mobility areas ζ_{R_x} intersect with every other, the system is *unconstrained*. In that case, and given the number of regions $|R|$, $2^{|R|}$ spatial configuration and system states are identified. Consequently, a *partly constrained* system is characterized by a range of 1 to $2^{|R|}$ system states.

In the documentation system, where the user and the race area are mobile in their respective areas $\big($Fig. $4(\zeta_{U_1})$ and $(\zeta_{S_1})\big)$, several configurations are identified (Fig. 5). These configurations take into account the P_1 and D_1 regions bounded intersection, and the impossible simultaneous intersection of U_1 with D_1 and S_1.

3.3 Spatial Configurations and Task Constraints

In the regatta documentation system, the entire scenario might be executed by a user located in the region P_1. However, as the user remains in the area ζ_{U_1}, only a part of the processing region is accessed $\big($Fig. $4(c)\big)$. Moreover, only five out of eleven spatial configurations characterize the situation of a user running a task in P_1 $\big($Fig. $5(b)\big)$.

In an adaptive prospect, every configuration should fall into a system state. In the documentation system, the nominal scenario and tasks are enriched with alternative system states built on top of three spatial configurations groups. When the user is in D_1 without being in P_1 $\big($Fig. $5(c)\big)$, the system provides a summary of the concepts ships and crews. The useful area covers both regions $\big($Fig. $4(d)\big)$. When the user is outside the component supported regions D_1 and P_1 $\big($Fig. $5(c)\big)$, she/he is provided with a minimap of the system coverage. At last, the user proximity to the racing area (*"when being close enough to the race area"* in the scenario) occurs when she/he stands in the race area, that is when $U_1 \cap S_1 \neq \emptyset$. $\big($Fig. $6(c)\big)$.

These spatial configurations become pre-requisites for tasks execution. They annotate tree nodes using their environmental requirements. For example, registering and rendering the user focus can be done when the system components spatial configuration belong to the group *(a)* in figure 5 $\big($Fig. $6(b)\big)$. When the user stands in the broadcasting region D_1, only the path leading to the subtask "Render static data" is executable $\big($Fig. $6(d)\big)$. In order to enable the task "take and share a picture", the user region has to intersect with the source region An

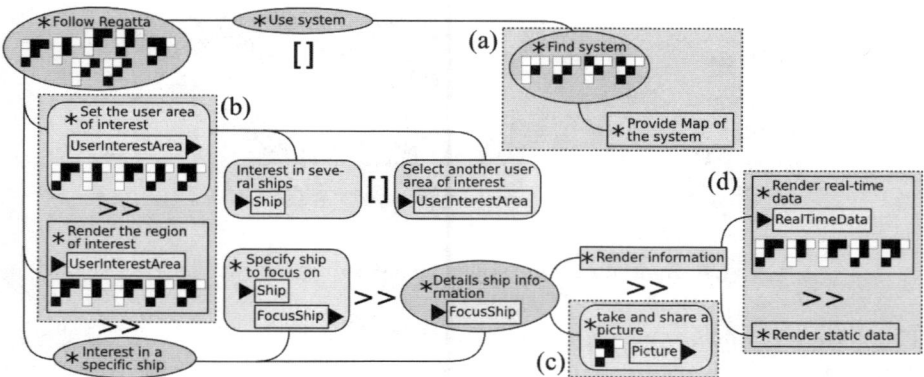

Fig. 6. Environment constrained tasks tree

unforeseen task completes the tree and is available when the user is outside all the regions (Fig. 6(a)). The abstract task, "Find system", is iteratively operated on the user platform. Reported at the top design level, this new task outlines an implicit alternative scenario to the nominal case.

4   Prototype Implementation

The task tree allows every system state to be part of a common user interface at the client level. Besides, the environmental conditions at the tree nodes level identify the procedure distribution. For example, the task "Provide map of the system" is enabled when the user stands outside of other regions. In that case, the task implementing procedures runs on the client platform.

The implementation of the proposed sailing race documentation system is still in progress. In order to give a brief overview of the intended results on the user platform, an illustrative walk of a user in the area of the system shows the relation between the system states and the functionality offered at the user-interface level (Fig. 7).

When being outside of any other region; the user is provided with a map of the system coverage(Fig. 7(a) - enabled task : "Provide a map of the system"). When the user comes in the broadcasting region at t_2, a sequential-access list of the ships shows-up (Fig. 7(b) - enabled task path : "Interest in a specificship" → "Render static data"). In the processing region, boats are mapped and their position are regularly derived. The user has an access to the detailed information by clicking on the displayed boats (Fig. 7(c) - this enables every tasks except "Take and share a picture"). Finally, when the sailing ships come near the shore-line, the user gets into the source region, and the task "take and share picture" is made accessible trough an enriched user interface (Fig. 7(d)). Given a small region available for services (Fig. 4(c)), the environment characterization has led to alternative system usages. Consequently, at the implementation stage, the useful area has been extended to cover entirely the user mobility area (Fig. 4(e)).

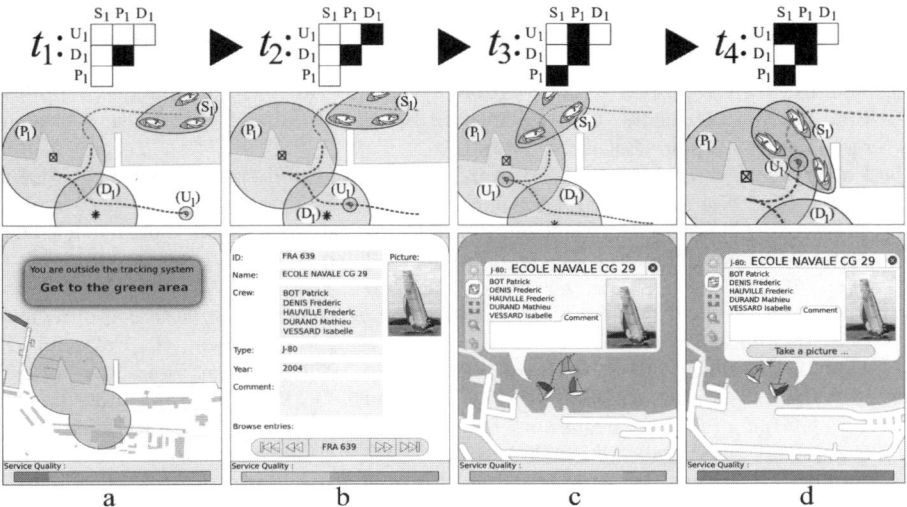

Fig. 7. Evolution of the system behavior in a dynamic environment

5 Conclusion

Coming along with growing interest in information technologies, progress made in ubiquitous computing and data management led GIS to the edge of interactive and information systems. While new usages of geographic information are emerging, novel design methodologies should be explored in order to move forward *geographic interactive systems*.

The research presented in this paper introduces a geographical extension to an interactive system design framework. From early conceptual stages, the spatial dimension of the environment is integrated within a user-centred design approach. The design process takes into account the runtime environment of the system. This enriches the nominal scenario by the generation of additional usages. The approach has been illustrated by the design of a distributed GIS, and work is still in progress regarding the validation in a real experimental context. Future work concerns the integration of the temporal dimension within the framework, and the design of the usage context at the user level.

References

1. Bastide, R., Navarre, D., Palanque, P.: A tool-supported design framework for safety critical interactive systems. Interacting with Computers 15(3), 309–328 (2003)
2. Bastien, J.M.C., Scapin, D.L.: Evaluating a user interface with ergonomic criteria. International Journal of Human-Computer Interaction 7(2), 105–121 (1995)

3. Baus, J., Krüger, A., Wahlster, W.: A resource-adaptive mobile navigation system. In: Proceedings of the 7th international conference on Intelligent user interfaces: IUI 2002, pp. 15–22 (2002)
4. Bédard, Y.: Geographical Information Systems: Principles, Techniques, Application and Managements. In: Principles of Spatial Database Analysis and Design, pp. 413–424. Wiley, Chichester (1999)
5. Boehm, B.: A spiral model of software development and enhancement. SIGSOFT Softw. Eng. Notes 11(4), 14–24 (1986)
6. Brown, B., Chalmers, M., MacColl, I.: Exploring tourism as a collaborative activity. Technical report, Equator-02-018 (September 2002),
   http://citeseer.ist.psu.edu/brown02exploring.html
7. Caffiau, S., Girard, P., Scapin, D., Guittet, L.: Generating interactive applications from task models: A hard challenge. In: Winckler, M., Johnson, H., Palanque, P. (eds.) TAMODIA 2007. LNCS, vol. 4849, pp. 267–272. Springer, Heidelberg (2007)
8. Calvary, G., Coutaz, J., Thevenin, D., Limbourg, Q., Bouillon, L., Vanderdonckt, J.: A unifying reference framework for multi-target user interfaces. Interacting with Computers 15(3), 289–308 (2003)
9. Hampe, M., Paelke, V.: Adaptive maps for mobile applications. In: Proceedings of the Mobile Maps Workshop at Mobile HCI 2005 (2005),
   http://www.ikg.uni-hannover.de/publikationen/publikationen/
10. Longley, P.A., Goodchild, M.F., Maguire, D.J., Rhind, D.W.: Geographic Information Systems and Sciences. 2nd edn. Wiley, Chichester (2005)
11. Luaces, M.R., Brisaboa, N.R., Paramá, J.R., Viqueira, J.R.: A generic framework for gis applications. In: Kwon, Y.-J., Bouju, A., Claramunt, C. (eds.) Proceedings of the 4th International Symposium on Web and Wireless Geographical Information Systems. pp. 94–109. Springer, Heidelberg (2004)
12. Mac Aoidh, E., Koinis, E., Bertolotto, M.: Improving archaeological heritage information access trough a personalized GIS interface. In: Carswell, J., Tekuza, T. (eds.) Proceedings of the 6th International Symposium on Web and Wireless Geographical Information Systems. pp. 135–145. Springer, Heidelberg (2006)
13. Mori, G., Paterno, F., Santoro, C.: CTTE: support for developing and analyzing task models for interactive system design. IEEE Transactions on Software Engineering 28(8), 797–813 (2002)
14. Parents, C., Spaccapietra, S., Zimànyi, E., Domini, P., Plazent, C., Vangenot, C.: Modeling Spatial Data in the MADS Conceptual Model. In: Proceedings of the 8th international symposium on Spatial Data Handling: SDH 1998, pp. 130–150 (1998)
15. Petit, M., Ray, C., Claramunt, C.: A contextual approach for the development of GIS: Application to maritime navigation. In: Carswell, J., Tekuza, T. (eds.) Proceedings of the 6th International Symposium on Web and Wireless Geographical Information Systems. pp. 158–169. Springer, Heidelberg (2006)
16. Rauschert, I., Agrawal, P., Sharma, R., Fuhrmann, S., Brewer, I., MacEachren, A.: Designing a human-centered, multimodal gis interface to support emergency management. In: Proceedings of the 10th ACM International Symposium on Advances in Geographic Information Systems. pp. 119–124. ACM Press, New York (2002)
17. Satyanarayanan, M.: Fundamental challenges in mobile computing. In: Proceedings of the fifteenth annual ACM symposium on Principles of distributed computing, pp. 1–7 (1996)
18. Yupin, Y., Shrum, L.J.: What is interactivity and is it always such a good thing? implications of definition, person, and situation for the influence of interactivity on advertising effectiveness. Journal of Advertising 31(4), 53–64 (2002)

Towards a Conceptual Model of Talking to
a Route Planner

Stephan Winter and Yunhui Wu

Department of Geomatics, The University of Melbourne, Victoria 3010, Australia
winter@unimelb.edu.au, y.wu21@pgrad.unimelb.edu.au

Abstract. Imagine a (web-based or mobile) route planning service that under-
stands and behaves like another person. This conceptual paper addresses the first
step towards this vision. It looks at the ways people would like to talk to their
route planner in the initial phase of the route communication when specifying
the travel route and time. The paper systematically collects service requirements,
based on elements from intelligent autonomous agents, and demonstrates that a
fundamental change is required compared to how services operate today.

Keywords: web-based route planning, human computer interface, ontologies,
place.

1 Introduction

This paper studies the common question of a person: "Can you tell me the way to ... ",
which initiates a typical everyday communication on routes, either with other persons,
or with dedicated route planning services[1]. Despite of its commonality this question
forms quite a challenge for route planning services. Looking at current services' appar-
ently simple interfaces—Figure 1 presents a typical subset of current route planners—it
appears that they fail to accept above's question as is, in its context and with all kinds
of location references a person may have in mind.

In another paper we have proposed a *spatial Turing test* for route planning services, postulating that services' behavior should aim to become indistinguishable from
human-generated route advice to overcome obstacles in their usage [53]. Going back
in that paper to Artificial Intelligence (in particular to [48]) was motivated by sharing
the general interests: On one hand the scientific interest in gaining knowledge of human
cognitive processes, and on the other hand the scientific interest in implementing formal
models of these processes in machines. It was also motivated by the observation that
communication with other people is more successful than with route planning services:

---

[1] For the purpose of this paper we include route planners of any form: web-based (desktop) or
wireless (mobile), local or centralized, off-line (planning before traveling) or on-line (accompanying the travel), and even printed ones (e.g., time tables), believing that the architecture is
relatively unimportant for our interest and also not always transparent to the wayfinder. We
also include route planners for any mode of traveling such as walking, cycling, driving, or
riding on public transport, trains or planes, but believe that in the future route planners have to
integrate modes of traveling.

M. Bertolotto, C. Ray, and X. Li (Eds.): W2GIS 2008, LNCS 5373, pp. 107–123, 2008.

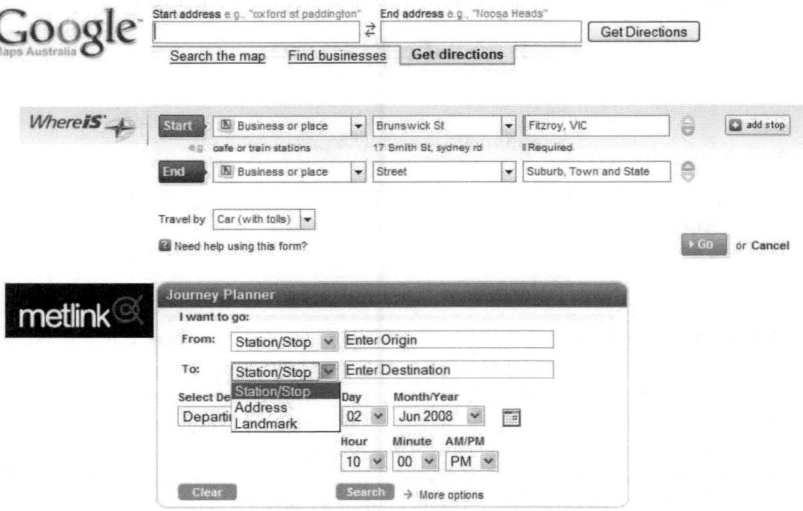

Fig. 1. The initial phase of a communication with different route planners (*Google Maps*, *WhereIs*, and Melbourne's public transport planner *metlink*)

- People have a richer spatial language, compared to the rigid syntax, grammar and ontology of route planners. But when talking to a route planning service, the interface requires the wayfinder to adapt to a restricted, rigid language (e.g., postal addresses, see Figure 1).
- People have more compatible spatial concepts and conceptualizations with each other, compared to the concepts provided by route planners. The concepts behind a route planner's interface are often determined by the service's internal needs, not by wayfinder needs (e.g., stop names of a bus network are difficult be known by wayfinders, see Figure 1).
- People have superior capabilities to capture a communication context and to adapt information to this context They communicate two-ways. Route planners have implemented this strategy only rudimentary (e.g., when resolving ambiguities by showing lists of alternatives) and cannot adapt.

Thus, route planners need to share some of the visions of Artificial Intelligence. They need to understand and support an autonomous and intelligent agent—the human wayfinder—although they are not autonomous by themselves. This means they should be able to cope with the characteristics and capabilities of an intelligent agent, which Brooks [1] has categorized as *situatedness*, *embodiment* and *emergence*. This paper will study the consequences of this postulate.

Winter and Wu [53] have looked at the complete dialog between a wayfinder and a route planner, from addressing the route planner to closing the communication. In contrast, the present paper will go a step further and focus on one element only, the initial phase where the wayfinder talks to the route planner to request a route. To illustrate anthropomorphic behavior in this phase let us consider the case of Mary. When one morning Mary jumps into her shoes to leave her apartment she would like to tell her device: "To uni!", and this device would tell her that she is in time to get the tram. This means the device would have succeeded to capture the context (from *here* and *now*, and *by public transport* is at least an option) and to match the colloquial and ambiguous location *uni*. Mary could also have asked her housemate sitting at her laptop about the way to the university: "Can you check my way to uni, please?" Well, the tone is different, but this question conveys the same content, and Mary's housemate would have had no problems either to understand her. But Mary, having an intelligent device—in Turing's sense of a device with anthropomorphic behavior—would not need to interrupt her housemate in her work.

This paper aims to develop the requirements of an anthropomorphic behavior for a service in the initial phase of a route communication. The hypothesis is that *the requirements lead to a fundamentally different wayfinder interface compared to current route planner interfaces*. If this is the case then route planning services cannot evolve into intelligent services; they require a paradigm shift in their internal design. To deal with the hypothesis we will study the three properties of an intelligent wayfinder: situatedness, embodiment and emergence. This will allow us to analyze the requirements of a service capable to understand their users.

To this extent this paper is purely conceptual, backed by research in spatial cognition. An experimental verification of the conceptual model by an implementation is beyond the scope of this paper and left for future work. Also left for future work are the consequences of this conceptual model for the other phases of the communication about routes.

The paper starts with a review of recent work (Section 2) to introduce the spatial Turing test. Then it studies the structure and tasks of the communication with a route planning service, in particular to identify the initial phase in which the wayfinder talks to the route planner (Section 3). Finally it embarks on the requirements analysis (Section 4) based on the desired characteristics identified in the previous section. The list of requirements is then discussed for the complexity of the challenge they pose (Section 5). This discussion shows strong evidence for the hypothesis. At the end the paper develops future research questions based on the requirements (Section 6).

2   Background

The Turing test of whether machines can think is an imitation game: a computer has to convince players that they are communicating with a person [48]. By this way Turing avoids to define *thinking* or *intelligence*; in Turing's sense a machine is intelligent if it shows anthropomorphic communication qualities.

We know that people reason on a hierarchical mental representation of space [17], which for instance leads to efficient communication about place [46]. We also know that a person's landmark-based, multi-granular route description is cognitively more efficient than a geometric route description as produced by current route planners. Human spatial representations are rather qualitative than geometric, and human communication of route directions nearly never refers to geometry in a metric manner [7,45] because people have difficulties to realize the metrics. Cognitive efficiency can be shown for example by navigational performance [5]. Research demonstrates that cognitively ergonomic route directions (for example, as given by persons) have some redundancy [32], have varying, if not asymmetric granularity [21,36,46], relate to the perceptual experience of the environment [6], and refer to a selected range of features depending on the traveling mode [2,24]. Current route planners cannot yet imitate persons in these ways. Accordingly, one aim of current research is generating landmark-based route descriptions automatically [12,15,31,33,46], which is surprisingly complex to realize in implementations.

Human route communication is yet superior to current services' route communication with respect to the cognitive effort of understanding, memorizing and realizing spatial descriptions in orientation and wayfinding. Hence, the spatial Turing test sets the bar for route planning services. Human communication has qualities challenging computers, such as adapting to the communication situation, or talking about places. The communication between a person and a route planner is further challenged because of the different fabric of internal spatial representations and reasoning in persons' minds [10,44] and in machines [39]. With different spatial concepts being involved, ontological mismatches will be expected in the communication of routes.

However, Turing held a philosophical interest on the nature of artificial intelligence. His test does not facilitate a real measure of intelligence of programs, nor does it provide an objective way of testing. For example, when comparing two spatial descriptions it is sometimes hard to say which one is *more* anthropomorphic. In this sense, we will be able to identify gaps leading to failure of the spatial Turing test, and based on these gaps develop requirements of a conceptual model of a route planner overcoming these gaps. But we will not be able to measure the quality of the communication with a route planner. This means, following Popper's research paradigm [30] the goal for this paper is to refute that machines cannot think spatially (by closing the gaps), but not to prove that machines can think spatially (i.e., prove that all gaps are closed).

3   Talking to a Route Planner

The communication between a wayfinder and a route planning service has the same structure of route communication between two persons. It consists of three phases: the initial phase where the wayfinder asks an informant for directions, the center phase where the informant provides route directions, and the closing phase of confirmations and separation [20,54]. Nearly all research in this area focuses on the center phase [6,19,34].

In this paper the focus is on the initial phase in which the wayfinder has the lead role and talks to the route planner. According to Klein [20, p. 168], the initial phase consists of three subtasks for the wayfinder:

- getting into contact with the informant;
- making clear what he wants;
- succeeding in getting the informant to take over the task of giving him route directions.

Orthogonally to the temporal sequence of tasks, Klein as well as Wunderlich and Reinelt [54, p. 183] identify three subtasks present at each stage of direction giving. These subtasks are a cognitive task (e.g., activating a spatial cognitive representation); an inter-actional task (e.g., initiating and terminating the verbal exchange, or providing a route description); and a linguistic task (e.g., expressing a comprehensible route description).

The initial phase was studied neither by Klein nor by Wunderlich and Reinelt in particular, and also other linguistic, cognitive and engineering research focuses rather on human or automatically generated route directions as a whole. However, it is possible to identify in the initial phase already:

- A cognitive task. The wayfinder has to find a proper specification for his route request, which means a specification that is sufficient for the informant in the given communication context. The informant has to activate a spatial cognitive representation and to identify the specification of the wayfinder in this representation.
- An interactional task. The wayfinder has to manage the three subtasks of the initial phase identified above. The informant has to pay attention, listen, and respond by confirming that the specification of a route was received and sufficient.
- A linguistic task. Wayfinder and informant interact via sign systems (a language), and all three subtasks of the initial phase have to be expressed in a sign system and understood by the recipient. The wayfinder has to contact the informant via language, express the request in a language, and be ensured via language that the informant took over.

In this paper the focus is on the interactions between the cognitive and the interactional task, which in the context of a machine as informant includes not only the cognitive abilities of the wayfinder, but also the internal data models and algorithms of the route planner. This means, the focus is on identifying and modeling the references that have to be conveyed and understood, and not on their actual representation in a specific sign system.

Technically, the three subtasks of the initial phase are dependent on the architecture and interface of the route planners. An example can be given for the three web-based route planners of Figure 1. The communication with a web-based planner is realized via the HTTP protocol. *Getting into contact* is realized by typing a uniform resource identifier in a web browser's address field, or alternatively by following hyperlinks to the route planners' home page (realizing an HTTP get request). *Making clear what the wayfinder wants* is dictated by the route planners' interface (Figure 1). *Succeeding in getting the route planner to take over* is realized by pressing a *submit* button (realizing another HTTP get request, parameterized by form data), and can prolong if there are ambiguities in the request to be resolved.

As outlined in Section 1 a machine, to pass the spatial Turing test, has to understand a wayfinder's request as another person would do. It was also argued above that a machine needs to have properties of an intelligent agent to do so. Hence, a route planner needs to be able to cope with *situatedness*, *embodiment* and *emergence* [1] as experienced by the wayfinder:

- Situatedness: the route planner should be aware of the context of the communication situation. This concerns in particular the location of the wayfinder and the time of the request.
- Embodiment: the route planner should be aware of the human intuitions underlying their ability to move and their common-sense or naïve understanding of the

world [10]. This concerns in particular relative, qualitative and egocentric spatial expressions.

- Emergence: the route planner should be aware of the coherent cognitive structures of the wayfinder evolved during the process of learning spatial environments. This concerns their procedural and declarative spatial knowledge, in particular the hierarchic organization of spatial cognitive representations.

The three elements will be used to structure the requirements analysis in the next section. To be precise, Brooks lists a fourth property of an intelligent agent, *intelligence*. In the present paper intelligence is seen emerging from the other three elements, applying here Turing's sense to the word, with no further need of a different definition and special requirements.

4 Requirements Analysis

What are the requirements of a route planner to pass the spatial Turing test with its initial communication phase? When Egenhofer identified database requirements for vehicle navigation in geographic space [9] he identified properties of data and operations wayfinders will want to perform. This wayfinder-centered approach will be continued here. This approach does not start from existing database properties or constraints, but from the behavior of wayfinders when making clear what they want, i.e., specifying a route request.

4.1 Situatedness

Each communication takes place in a situation, which forms the context of the communication. The context provides the multiple perspectives needed for understanding the discourse. While people are good at capturing the context, for machines such as route planning services this is a challenge. This section develops a conceptual model of the context of route communication.

According to a classification by Janelle [18], spatial constraints of a communication situation are either *physical co-presence* or *telepresence*, while temporal constraints are either *synchronous* or *asynchronous* communication. The cases are discussed in Table 1.

Interestingly, in the context of seeking route advice this categorization is insufficient. To capture the full spatiotemporal communication context one needs also to consider the spatiotemporal constraints of the activity the communication is about: wayfinding. One of the two spatial constraints of this activity—the point of departure—can be the current location of the wayfinder (*from here*), but the communication can also concern a future

Table 1. Janelle's spatial and temporal communication constraints [18] applied on seeking route advice

	synchronous	*asynchronous*
physical co-presence	e.g., face-to-face, or from mobile location-aware device	e.g., from you-are-here maps, or departure plans at bus stops
telepresence	e.g., via telephone or from web service	e.g., departure plan from a web page

activity with any other point of departure (*from elsewhere*). The other spatial constraint of the activity, the desired destination, is independent from any communication context, hence, needs to be specified in any case, and is excluded here from consideration. The temporal constraint of the activity—the time of departure—can be the actual time of the communication (*now*), but it can also be for any departure in the future (*at a later time*). These constraints are not completely orthogonal; if a request is for *now* it is necessarily for *here* (but not vice versa). Also, sometimes it is not the time of departure but the time of arrival that shall be specified. This case is equivalent to any departure in the future since it is not specific about the time of departure. Together these dimensions span a four-dimensional space (compared to the two-dimensional space of Table 1), with two categories each:

- location of the communicators: *physical co-presence* or *telepresence*
- time of the communication: *synchronous* or *asynchronous*
- point of departure: *from here* or *from elsewhere*
- time of departure: *now* or *at a later time*

Let us call any four-tuple of these parameters the *context* of the communication. Some typical projections of this space to a specific context are:

- <*physical co-presence, synchronous, from here, now*>:
  - A person asking a passer-by: "Can you tell me the way to the city hall?"
  - A person studying a bus stop's real-time arrival and departure display.
  - A tourist using a mobile location-aware device as a guide for a city tour.
- <*physical co-presence, asynchronous, from here, at a later time*>:
  - A car driver is following the signs to the airport.
  - A traveler studying a train station's departure plan.

It is left open whether all 16 possible combinations of context parameters have a reflection in communication scenarios; the observation of a correlation between *now* and *here* runs counter such an expectation. The latter case indicates already that the perceived or assumed context is not necessarily identical for the wayfinder and a service. Some specific context mismatches are for example:

- A public transport user arriving at a tram stop is searching for the next tram going to city (<*physical co-presence, synchronous, from here, now*>) and finds a plan of departure times (<*physical co-presence, asynchronous, from here, at a later time*>).
- A public transport user, just before leaving home, is turning to his desktop for advice on a trip across the city (<*telepresence, synchronous, from here, now*>) and finds a service interface asking for a departure location and time (<. . . , *from elsewhere, at a later time*>).

A clear condition for a service to pass the spatial Turing test is to be able to adapt to the context of the wayfinder to an extent any person being asked for route advice would do (see the example above, or [14,20,54]). Thus, the first postulated requirements are:

1. A service should be accessible anywhere, anytime [49].

   When a wayfinding problem arises persons can not solve on their own, they will ask other persons around for advice. A mobile device or a terminal has the same affordance. While the wayfinder does not care whether a service is running locally (*physical co-presence*) or centrally (*telepresence*), the wayfinder needs local access, and in case of centralized service architectures this means ubiquitous connectivity.

2. A service should be able to anticipate the spatiotemporal activity context from the communication situation.

   If a communication situation suggests a request for *from here*, this context specification *from here* should be inferred automatically. Face-to-face communication between people (*physical co-presence*) does without explicit communication of *from here*. In contrast, any *telepresence* communication situation between people will make wayfinders specify their current location on their own initiative. The problem is that while a wayfinder calling a person knows that this other person has no access to the caller's current location and adapts to this situation, this wayfinder cannot recognize whether a service accessible from a local device is running locally (*physical co-presence*) or centralized (*telepresence*). Hence, the situation's affordance does not trigger wayfinders to reveal their current location.

   Furthermore, communication between people does without explicit communication of *now* if it can be concluded from the communication situation. Services should be able to behave similarly.

3. A service should be able to understand the human cognitive concepts that are used in the communication about the activity context. This concerns in particular the concepts of *from here* and *now*, but also *from elsewhere* and *at a later time*.

   Route planning services take various inputs, including origin, destination and time of travel, to produce first a route and then a route description. This input needs to be understood by the service such that it can derive a complete parametrization for the optimal route algorithm. But this input should also have anthropomorphic properties to pass a spatial Turing test, i.e., it should allow wayfinders to communicate in their terms. And it is the services's obligation to do the transfer.

The third requirement has consequences for embodiment and emergence, and hence, the discussion re-appears in the following Sections 4.2 and 4.3.

4.2   Embodiment

Embodiment in this context postulates a route planner to consider that the wayfinder is a subject having and using a body for movement, and these sensorimotor skills and experiences shape the person's spatial and temporal cognitive concepts. In the center of the initial phase of communication is the specification of a change of place by the wayfinder. The route planning service, to pass a spatial Turing test, should allow wayfinders to communicate their specification in their spatial and temporal terms. Besides of considering the situation (Section 4.1), this means to accept and interpret the spatial and temporal references in the verbal expression of the wayfinder.

People answer *where* questions in a hierarchic manner, either zooming in or zooming out [28,29,37,38]. "San Francisco is in the USA, in California." – "My office is in the

Department of Geomatics, which is part of the School of Engineering at the university." – "The newspaper is on the table in the living room." – Place descriptions of this form are a special case of general *referring expressions* [3]. They are not only hierarchically organized, they are also finite (and even relatively short) despite the infinite possibilities to describe a place. A speaker includes just the references that are sufficient in the given communication context. This context is defined by the location where the communication takes place (*here*) and the place to be described. In fact, initializing the route communication is one form of answering *where* questions ("Where are you?", "Where do you want to go?").

If the location of the communication is close to the location to be described, references to more global places are neglected, and references to more detailed places are added:

– Conversation at home: "The newspaper is on the table in the living room [of our apartment, which is in Melbourne, Australia]."
– Conversation in a coffee shop: "I left the newspaper [in Australia, in Melbourne,] at home [in the living room, on the table].")

Similarly, but more on a geographic scale:

– Conversation overseas: "I am living in Melbourne, Australia."
– Conversation when entering a taxi at Melbourne airport: "To North Melbourne, near the hospitals, off Flemington road."
– Conversation when friends are ringing the intercom: "Third floor, the apartment opposite the lift."

For the speaker different references appear to be relevant in different contexts. Applying relevance theory [51] we can already determine automatically which references are relevant, based on the distance of references in a hierarchical conceptualization of the environment [4,26,27,46]. Current route planners do not. First, they have a prescriptive and static interface. Secondly, if toponym resolution leads into ambiguities (Melbourne in Australia or in Florida, USA?) some planners always insist that the wayfinder resolves the ambiguity, while other ones at least interpret the destination location in the context of the given start location (if the first address is in Melbourne, Australia, the second address is assumed by default to be in Melbourne, Australia, even if the same street name exist in Melbourne, Florida).

What has been discussed for space so far extends to time in the same manner. Temporal references have a granularity ("tomorrow", "tomorrow afternoon", "tomorrow at 2pm", "tomorrow at 2:18pm"), and which granularity is relevant is determined by the communication context.

Furthermore, place descriptions come with spatial prepositions denoting spatial relations between subject and object (see the examples above). Understanding spatial prepositions is a challenge for natural language understanding [42], and current web services—route planners as well as local search—interpret only the most basic spatial relationship: *in*, ignoring all others (except a couple of services interpreting the binary relationship *corner*). Understanding is impeded by the context dependency and vagueness of qualitative relationships such as *near* [8].

The second set of requirements, extracted from this discussion, is therefore:

4. A service should always be aware of the current location of the wayfinder, enabling the service for accepting deictic references.
5. A service should apply concepts of relevance to the route specification.
6. A service should have access to hierarchic representations of space and time, and be able to navigate in these representations.
7. A service should have capabilities to understand spatial prepositions and relations.

Related to the embodied experience of space and time is the structure of the immediate environment of the wayfinder. However, this aspect is covered by the discussion of emergence in the next section.

4.3   Emergence

Emergence in complex systems refers to the way patterns at macroscopic levels arise out of a large number of interactions at microscopic levels. In the current context the relevant complex system is the procedural and declarative knowledge of humans of their environment. It was mentioned already that cognitive spatial representations have a hierarchical order. Accordingly, human reasoning for route planning is hierarchically organized, which has inspired hierarchical route planning strategies [43,50]. Movement is planned at higher levels first, and then filled with actions at lower levels (down to the subconscious level of body coordination during locomotion). Movement specifications, as they contain the description of the destination, can be hierarchic as well (see Section 4.2).

An aspect not discussed before are the structures and constraints of the environment of the wayfinder at different levels of these hierarchies. Of central interest are the human concepts of *here* and *now*, both having vague and context-dependent semantics. *Here* and *now* may be references to places and times over some levels of granularity, and at each level place and time may not be crisply defined. The point of departure being *here* can mean the actual position of the wayfinder, but also generalized concepts of place, or the nearest access to a transportation network. Similarly, the departure *now* can mean immediate departure but also departure with the next available transportation means, or the one after if its more convenient.

Imagine planning to travel from Melbourne to Tokyo. Current route planners are based on the airline transportation networks[2]. Automatically they identify places such as *Melbourne* or *Tokyo* with airports, as airports represent the nodes in these transportation networks. They expect that the wayfinder is able to find the way to the airport in Melbourne on their own, or alternatively chooses another route planner for this part of the travel. The public transport planner for Melbourne[3] might help here. It guides the wayfinder from a nearby tram stop to the airport, using the airport bus. The bus arrives in front of the arrival hall, but the planner tells the wayfinder that from the arrival stop a walk of 230m leads to *Melbourne Airport*. Whatever the granularity of this metric detail is, this destination is not the wayfinder's check-in counter. Thus, the wayfinder

---

[2] For example, http://www.quantas.com or http://www.expedia.com
[3] http://metlinkmelbourne.com.au

needs another route planner for finding the way through the airport. This time she might choose a display board of departures, or the information booth, or a You-Are-Here map. To find one of these she has to make route planning decisions on her own, based on perceived affordances. – Combining these individual route planning processes to one, one finds a recursively refining specification of the route: "I want to go [from here] to Tokyo." – "Ok, but tell me how to find to the airport." – "And at the airport, where is the check-in counter for the flight to Tokyo?" And so on. These multiple hierarchies in route planning and communication were studied, for example, by Rüetschi and Timpf [36]. Current route planners fail to realize this properly.

Now let us assume that the wayfinder has specified in all cases the time of departure as *now* (or a time in the future). The airline route planner knows only that the wayfinder is *in Melbourne*. The public transport planner ignores that she is at home at the time of the request (or she ignores to specify in her request that she is currently at home). Only the display board of departures at the tram stop assumes that the wayfinder is within eyespot (strictest *physical co-presence*). With other words, the notions of *here* and *now* have different meanings in different communication contexts (again, this context is determined by the current location in comparison to the destination, which varies here between *Tokyo*, the *airport* and the *check-in counter*).

Thus, the following requirements can be derived from the discussion:

8. A service should capture the spatiotemporal constraints of the route to resolve the vagueness of wayfinder-defined terms *here* and *now*.
9. A service should apply means of time geography [25] at all levels of spatial and temporal granularity to deal with the inherent vagueness of *here* and *now*.
10. A service should, therefore, acknowledge that all route planning is inherently time-dependent. People may give different route directions by day than by night, or at peak hours or non-peak hours.
11. A service should care for integrated route planning on different transportation networks (of different scales and different means of transportation).

In principle, also the spatial and temporal constraints of the communication come with vaguely defined categories. *Physical co-presence* may extend, for example, to earshot or eyespot, and *telepresence* as its counterpart has complementary vague meaning. *Synchronous* communication may accept some delay such as we experienced in the early days of Skype[4]. However, the communication context is not verbally communicated. Thus the task of understanding and mapping human notions is limited to the spatiotemporal constraints of the activity, *here* and *now*.

5   Comparison of the Requirements with Current Route Planning Services

In this micro-cosmos of a person talking to a route planner at least two ontologies (or epistemologies, to be precise) [40,13] meet: the human wayfinder's ontology of the environment and the service's internal ontology. Talking to a route planner a wayfinder

---

[4] Skype is a registered trademark.

provides place descriptions, at least for the destination, and potentially a time description. These descriptions have to be mapped by the route planning service to its internal ontology. Essentially this forms a problem of *understanding* place descriptions.

The service's internal ontology is defined by graph theory and network analysis. To determine a route it needs to translate any input to two unambiguous positions on a transportation network, the start s' and the destination d'. This process of translation may include toponym identification and resolution, georeferencing the toponyms by looking up gazetteers [41], map matching to draw the georeference to the next point on the transportation network, and dynamic segmentation of the transportation network to include temporarily these two points to the network for a shortest path algorithm. Additionally, for time-dependent shortest path problems one of both times is required: the departure time or the arrival time t'. This internal ontology of a service, (s', d', t'), shows in any current route planner interface (e.g., Figure 1). These interfaces are static and do not adopt to the situation, except that they may fill the form fields with default values.

In current route planners this translation has restricted functionality. While gazetteers help to resolve toponyms, current gazetteers contain only authoritative geographic placenames [16]. Other databases may add privately collected placenames such as the entries in Yellow Pages, or the stop names of a public transport provider. However, all these resources do not yet contain synonyms, vernacular placenames or common abbreviations of placenames. They also do not contain information about the spatial extent of a place, or of its relationships with other places. Hence, route planners are typically structuring and limiting the input by a wayfinder. Furthermore, most if not all spatial prepositions and relationships cannot be resolved and are ignored.

In comparison the wayfinder ontology is loosely captured by the requirements listed in Section 4. The 11 requirements go beyond the capabilities of current route planners. To prove the hypothesis we have to demonstrate that these requirements demand fundamental changes to the interface as well as the internal functionality of current route planners.

In this respect, not all requirements have the same standing. Some requirements look at fundamental change for some route planners. An example is R10 ("Acknowledge that all route planning is inherently time-dependent"). Typical route planning services for walking, biking or car navigation do not consider time. To match better anthropomorphic behavior they should. But the knowledge to realize these fundamental changes exists, and some services comply already, such as vehicle navigation systems linking with traffic message channel. Thus, requirements of this sort do not prove the hypothesis.

But other requirements stand out. The most challenging ones may be R3, R6 and R7, or understanding the human concepts of *here* and *now*, hierarchical place descriptions, and spatial prepositions. All three of them are in conflict with current interfaces as well as with current capacities of language understanding. At least with these three requirements the hypothesis is proven.

6   Conclusions

This paper studies the initial phase of the communication between a human wayfinder and a route planning service. It starts from requirements derived by a vision of a service of anthropomorphic understanding and behavior. These requirements are fundamentally different from the ontologies currently implemented, and hence, require a new approach to accept and understand wayfinders' expressions about their spatial and temporal travel constraints. In particular we argue that a service needs to be able to understand place descriptions, human concepts of *here* and *now*, and the spatial prepositions in natural language expressions.

The approach to derive these requirements was based on the argument that a route planning service deals with an autonomous and mobile physical agent, and hence, has to reflect the elements of this agent: being situated, embodied and capable of dealing with a complex geographic environment. Although this approach is systematic, no claims are made to be complete. The identified requirements will close some gaps on the way to a service that would pass a spatial Turing test.

Some of the remaining open questions concern:

- The construction of cognitively motivated hierarchic structures of spatial representations. To reflect the cognitive representations of people of their environment, these hierarchies should contain synonyms, elements of vernacular geography (where boundaries are often different to authoritative boundaries) and for all elements of a city [23]. Accordingly, toponyms should be represented by their spatial extent, and this requires a capacity to represent and deal with vague concepts. Furthermore, the order in the hierarchy should be determined by cognitively salient characteristics. Some research in this direction has started [47,52], but a consistent, comprehensive and integrated model for all elements of a city or any other environment is missing.
- Free-form place descriptions consist not only of hierarchically ordered toponyms but also of spatial prepositions that need to be interpreted to fully capture the sense of a place description. While the hierarchy represents *part-of* relationships, other spatial relationships between geographic features need to be represented as well, implicitly or explicitly. Some of them are already captured in spatial databases, for example on connectivity and topology. But since they are not (yet) available in gazetteers, methods to exploit them for understanding place descriptions are lacking. What is also lacking are models for the interpretation of relations of vague or context-dependent meaning, such as *nearness* [8]. Going further, Kuhn postulates research in cognitive semantic engineering [22]. The latter establishes models capturing the semantics of spatial concepts, acknowledging that perception (which informs human conceptualization) is situated and embodied, and Gestalt is emerging from complex perceptive and cognitive processes in a complex spatial environment.
- Since understanding place descriptions is a special case of natural language understanding, the vertical and horizontal relationships captured before have to be applied for the understanding of human place descriptions. While the preliminary hierarchies are successfully used to automatically generate place descriptions [35,46], the reverse process of understanding given place descriptions is not yet investigated. Again, an area where we can profit from current research is the interpretation of sketches in graphical query interfaces to geographic information systems [11].

Although in this particular case the sign system is different from natural[5] language, the subject is the same: communication about place.

Furthermore, within the given context of a route communication we so far expected descriptions of enduring places in geographic space, such that toponyms can be stored in databases. However, more transient cases of place exist as well, and some of them can even have a meaning in route communication. A person may, for example, look for the way to a meeting. And metaphorical places (e.g., heaven) will further be excluded.

Acknowledgments

Part of this work is supported under the Australian Research Council's Discovery Projects funding scheme (project number 0878119).

References

1. Brooks, R.A.: Intelligence without reason. In: Myopoulos, J., Reiter, R. (eds.) 12th International Joint Conference on Artificial Intelligence IJCAI 1991. San Mateo, CA, pp. 569–595. Morgan Kaufmann Publishers, San Francisco (1991)
2. Burnett, G.E.: "Turn right at the traffic lights": The requirement for landmarks in vehicle navigation systems. The Journal of Navigation 53(3), 499–510 (2000)
3. Dale, R.: Generating Referring Expressions: Constructing Descriptions in a Domain of Objects and Processes. MIT Press, Cambridge (1992)
4. Dale, R., Geldof, S., Prost, J.P.: Using Natural Language Generation in Automatic Route Description. Journal of Research and Practice in Information Technology 37(1), 89–105 (2005)
5. Daniel, M.P., Tom, A., Manghi, E., Denis, M.: Testing the value of route directions through navigational performance. Spatial Cognition and Computation 3(4), 269–289 (2004)
6. Denis, M.: The description of routes: A cognitive approach to the production of spatial discourse. Current Psychology of Cognition 16, 409–458 (1997)
7. Denis, M., Pazzaglia, F., Cornoldi, C., Bertolo, L.: Spatial Discourse and Navigation: An Analysis of Route Directions in the City of Venice. Applied Cognitive Psychology 13, 145–174 (1999)
8. Duckham, M., Worboys, M.: Computational Structure in Three-Valued Nearness Relations. In: Montello, D.R. (ed.) COSIT 2001. LNCS, vol. 2205, pp. 76–91. Springer, Heidelberg (2001)
9. Egenhofer, M.: What's special about spatial? Database requirements for vehicle navigation in geographic space. In: ACM SIGMOD International Conference on Management of Data, Washington, pp. 398–402. ACM Press, New York (1993)
10. Egenhofer, M.J., Mark, D.M.: Naive Geography. In: Frank, A.U., Kuhn, W. (eds.) Spatial Information Theory. LNCS, vol. 988, pp. 1–15. Springer, Berlin (1995)
11. Egenhofer, M.J.: Query Processing in Spatial-Query-by-Sketch. Journal of Visual Languages and Computing 8(4), 403–424 (1997)

---

[5] Although this might be the correct technical term to characterize spoken or written discourse, one could argue that drawing sketches is also a *natural* way of expression. It is probably even the older way to communicate than by words.

12. Elias, B., Sester, M.: Incorporating landmarks with quality measures in routing procedures. In: Raubal, M., Miller, H.J., Frank, A.U., Goodchild, M.F. (eds.) GIScience 2006. LNCS, vol. 4197, pp. 65–80. Springer, Heidelberg (2006)

13. Frank, A.U.: Ontology. In: Kemp, K.K. (ed.) Encyclopedia of Geographic Information Science. Sage Publications, Thousand Oaks (2007)

14. Freundschuh, S.M., Mark, D.M., Gopal, S., Gould, M.D., Couclelis, H.: Verbal Directions for Wayfinding: Implications for Navigation and Geographic Information and Analysis Systems. In: Brassel, K., Kishimoto, H. (eds.) 4th International Symposium on Spatial Data Handling, Zurich, Department of Geography, pp. 478–487. University of Zurich (1990)

15. Hansen, S., Richter, K.F., Klippel, A.: Landmarks in OpenLS: A Data Structure for Cognitive Ergonomic Route Directions. In: Raubal, M., Miller, H.J., Frank, A.U., Goodchild, M.F. (eds.) GIScience 2006. LNCS, vol. 4197, pp. 128–144. Springer, Heidelberg (2006)

16. Hill, L.L.: Georeferencing: The Geographic Associations of Information. In: Digital Libraries and Electronic Publishing. MIT Press, Cambridge (2006)

17. Hirtle, S.C., Jonides, J.: Evidence of hierarchies in cognitive maps. Memory and Cognition 13(3), 208–217 (1985)

18. Janelle, D.G.: Impact of information technologies. In: Hanson, S., Giuliano, G. (eds.) The Geography of Urban Transportation. pp. 86–112. Guilford Press, New York (2004)

19. Klein, W.: Wegauskünfte. Zeitschrift für Literaturwissenschaft und Linguistik 33, 9–57 (1979)

20. Klein, W.: Local Deixis in Route Directions. In: Jarvella, R.J., Klein, W. (eds.) Speech, Place, and Action. pp. 161–182. John Wiley & Sons, Chichester (1982)

21. Klippel, A., Tappe, H., Habel, C.: Pictorial representations of routes: Chunking route segments during comprehension. In: Freksa, C., Brauer, W., Habel, C., Wender, K.F. (eds.) Spatial Cognition III. LNCS, vol. 2685, pp. 11–33. Springer, Berlin (2003)

22. Kuhn, W.: Why Information Science needs Cognitive Semantics. In: Workshop on the Potential of Cognitive Semantics for Ontologies (FOIS 2004), Torino, Italy (2004)

23. Lynch, K.: The Image of the City. MIT Press, Cambridge (1960)

24. Maaß, W.: Von visuellen Daten zu inkrementellen Wegbeschreibungen in dreidimensionalen Umgebungen: Das Modell eines kognitiven Agenten. Phd thesis, Universität des Saarlandes (1996)

25. Miller, H.J.: Activities in Space and Time. In: Hensher, D.A., Button, K.J., Haynes, K.E., Stopher, P.R. (eds.) Handbook of Transport Geography and Spatial Systems. Handbooks in Transport, vol. 5, pp. 647–660. Elsevier, Amsterdam (2004)

26. Paraboni, I., Deemter, K.: Generating Easy References: The Case of Document Deixis. In: Second International Conference on Natural Language Generation (INLG 2002), New York, USA, pp. 113–119 (2002)

27. Paraboni, I., Deemter, K.v., Masthoff, J.: Generating referring expressions: Making referents easy to identify. Computational Linguistics 33(2), 229–254 (2007)

28. Plumert, J.M., Carswell, C., DeVet, K., Ihrig, D.: The Content and Organization of Communication about Object Locations. Journal of Memory and Language 34, 477–498 (1995)

29. Plumert, J.M., Spalding, T.L., Nichols-Whitehead, P.: Preferences for ascending and descending hierarchical organization in spatial communication. Memory and Cognition 29(2), 274–284 (2001)

30. Popper, K.: The Logic of Scientific Discovery. Routledge Classics. Routledge, London (2002)

31. Raubal, M., Winter, S.: Enriching Wayfinding Instructions with Local Landmarks. In: Egenhofer, M.J., Mark, D.M. (eds.) GIScience 2002. LNCS, vol. 2478, pp. 243–259. Springer, Berlin (2002)

32. Richter, K.F., Klippel, A.: A Model for Context-Specific Route Directions. In: Freksa, C., Knauff, M., Krieg-Brückner, B., Nebel, B., Barkowsky, T. (eds.) Spatial Cognition IV. LNCS, vol. 3343, pp. 58–78. Springer, Berlin (2005)

33. Richter, K.F., Klippel, A.: Before or after: Prepositions in spatially constrained systems. In: Barkowsky, T., Knauff, M., Ligozat, G., Montello, D.R. (eds.) Spatial Cognition 2007. LNCS, vol. 4387, pp. 453–469. Springer, Berlin (2007)

34. Richter, K.F.: Context-Specific Route Directions. Monograph Series of the Transregional Collaborative Research Center SFB/TR8, vol. 3. Akademische Verlagsgesellschaft, Berlin (2008)

35. Richter, K.F., Tomko, M., Winter, S.: A dialog-driven process of generating route directions. Computers, Environment and Urban Systems 32(3), 233–245 (2008)

36. Rüetschi, U.J., Timpf, S.: Modelling wayfinding in public transport: Network space and scene space. In: Freksa, C., Knauff, M., Krieg-Brückner, B., Nebel, B., Barkowsky, T. (eds.) Spatial Cognition IV. LNCS, vol. 3343, pp. 24–41. Springer, Berlin (2005)

37. Shanon, B.: Where Questions (1979), http://acl.ldc.upenn.edu//P/P79/P79-1017.pdf

38. Shanon, B.: Answers to Where-Questions. Discourse Processes 6, 319–352 (1983)

39. Shekhar, S., Vatsavai, R.R., Ma, X., Yoo, J.S.: Navigation systems: A spatial database perspective. In: Schiller, J., Voisard, A. (eds.) Location-Based Services, pp. 41–80. Morgan Kaufmann Pubblishers, San Francisco (2004)

40. Smith, B.: Ontology and information science. In: Zalta, E.N., Nodelman, U., Allen, C. (eds.) Stanford Encyclopedia of Philosophy. Center for the Study of Language and Information, Stanford University, Stanford (2003)

41. Stokes, N., Li, Y., Moffat, A., Rong, J.: An empirical study of the effects of NLP components on geographic IR performance. International Journal of Geographical Information Science 22(3), 247–264 (2008)

42. Tenbrink, T.: Space, Time, and the Use of Language: An Investigation of Relationships. Mouton de Gruyter, Berlin (2007)

43. Timpf, S., Volta, G.S., Pollock, D.W., Frank, A.U., Egenhofer, M.J.: A conceptual model of wayfinding using multiple levels of abstraction. In: Frank, A.U., Campari, I., Formentini, U. (eds.) GIS 1992. LNCS, vol. 639, pp. 348–367. Springer, Berlin (1992)

44. Timpf, S.: Ontologies of wayfinding: A traveler's perspective. Networks and Spatial Economics 2(1), 9–33 (2002)

45. Tom, A., Denis, M.: Referring to landmark or street information in route directions: What difference does it make? In: Kuhn, W., Worboys, M.F., Timpf, S. (eds.) COSIT 2003. LNCS, vol. 2825, pp. 362–374. Springer, Berlin (2003)

46. Tomko, M., Winter, S.: Pragmatic construction of destination descriptions for urban environments. Spatial Cognition and Computation (accepted August 21, 2008)

47. Tomko, M., Winter, S., Claramunt, C.: Experiential hierarchies of streets. Computers, Environment and Urban Systems 32(1), 41–52 (2008)

48. Turing, A.M.: Computing machinery and intelligence. Mind 59(236), 433–460 (1950)

49. Weiser, M.: The Computer for the Twenty-First Century. Scientific American (9), 94–104 (1991)

50. Wiener, J.M., Mallot, H.A.: Fine-to-coarse route planning and navigation in regionalized environments. Spatial Cognition and Computation 3(4), 331–358 (2003)

51. Wilson, D., Sperber, D.: Relevance Theory. In: Horn, L.R., Ward, G. (eds.) Handbook of Pragmatics, pp. 607–632. Blackwell, Oxford (2004)
52. Winter, S., Tomko, M., Elias, B., Sester, M.: Landmark hierarchies in context. Environment and Planning B 35(3), 381–398 (2008)
53. Winter, S., Wu, Y.: The spatial Turing test. In: Navratil, G. (ed.) Colloquium for Andrew U. Frank's 60th Birthday, Geoinfo Series, Vienna, Austria. vol. 39, pp. 109–116. Department for Geoinformation and Cartography, Technical University Vienna (2008)
54. Wunderlich, D., Reinelt, R.: How to get there from here. In: Jarvella, R.J., Klein, W. (eds.) Speech, Place, and Action, pp. 183–201. John Wiley & Sons, Chichester (1982)

Estimation of Geographic Relevance for Web Objects Using Probabilistic Models

Taro Tezuka[1], Hiroyuki Kondo[2], and Katsumi Tanaka[2]

[1] Institute of Science and Engineering, Ritsumeikan University
1-1-1 Noji-higashi, Kusatsu-shi, Shiga, 525-8577, Japan
`tezuka@media.ritsumei.ac.jp`
[2] Graduate School of Informatics, Kyoto University
Yoshida-honmachi, Sakyo-ku, Kyoto, 606-8501, Japan
`{kondo,tanaka}@dl.kuis.kyoto-u.ac.jp`

Abstract. The rapidly increasing use of geographically restrictive web search has made determination of the geographic relevance of web content an important task. We have developed a method that uses Gaussian mixture models to estimate the geographic relevance of web pages and arbitrary topics and have implemented a visualization interface that maps pages and topics to geographic space. The system enables the user to retrieve web pages and topics expressed on the Web that are relevant to an arbitrary geographic area.

Keywords: Geographic web search, Gaussian mixture models, Topic-level search.

1 Introduction

Geographic (or "local") web searches are among the most widely used web applications. There are a number of implementations, some available through prominent providers of web search engines [29,30,31], and a crucial task in these applications is the mapping of web pages to geographic space. The *geographic relevance* of a web page must be estimated in addition to the conventional relevance measures used in information retrieval [1].

In this paper we show how probability density functions can be used to model the geographic relevance of *web objects*. We use the term "web object" here to refer to both a web page and a collection of pages. For example, if a query is "Which city in New England is famous for autumn leaves?", our system returns a list of the locations most relevant to autumn leaves in New England so that users can get geographically restricted information more efficiently than they can by looking through all web pages relevant to autumn leaves.

Our system gathers web pages relevant to the user's query and estimates their geographic relevance by analyzing their contents. We wanted to build a real-time application using web services, rather than one indexing massive crawled data by preprocessing. Since the number of documents that can be collected in real time is limited, their geographic relevance must be estimated using a small number of texts that might may not contain addresses. We have therefore developed a method that estimates geographic relevance from place names in general instead of using addresses, as many other implementations of geographically restricted searches do.

M. Bertolotto, C. Ray, and X. Li (Eds.): W2GIS 2008, LNCS 5373, pp. 124–139, 2008.

One problem that occurs when using place names in general is that they often have ambiguity. The system is therefore required to make a good estimation by filtering out irrelevant coordinates. We show here the results of experiments demonstrating that this problem can be dealt with effectively by using Gaussian mixture models.

Using our system, users can retrieve web pages and topics expressed on the Web that are relevant to an arbitrary area, and fulfill their needs for geographic information more effectively.

The rest of the paper is organized as follows. Section 2 describes related work, Section 3 explains our model of geographic relevance, and Section 4 describes our method for estimating that relevance. Section 5 describes our implementation, Section 6 presents the results we obtained when evaluating that implementation, and Section 7 concludes the paper by summarizing it briefly and indicating directions for future work.

2   Related Work

In this section, we discuss some existing work on page-level and topic-level geographic web search systems that are related to ours. We first discuss geographic web search in general, and then describe more fundamental technologies that are used in our research.

2.1   Geographic Web Search

A classic work by McCurley mapped web pages to geographic locations by using IP addresses, physical addresses, telephone numbers, and other information [2]. Although a set of geographical information is presented to the user as he browse, the system does not perform any information aggregation technology to extract new information which is not explicitly expressed on the Web.

Gao et al. implemented a system that utilizes location information of web pages to collect web pages evenly from different regions [3]. Their method helps build web search engines with rich geographic information, yet they do not go as far as extracting useful geographic information from their data repository.

Mei et al. extracted spatial and temporal distributions of topics using Weblogs [13]. Their approach is very much similar to ours. Geographical scale that they have investigated, however, is at a national level, comparing differences in frequency of terms between states. Our system enables the user to choose the scale at which the estimation of the distribution is to be made, through the map interface. Also, their research was primarily on investigating whether such geographic bias would exists, while we aim at building a practical system, presenting the result of the estimation to the user in real time.

Tezuka et al. presented various new applications that integrates the Web and Geographic Information Systems[14]. Applications included extraction of geographically significant objects using the Web, and integration of personal experiences expressed on Weblogs.

Gravano et al. used support vector machines to classify queries sent to web search engines as local or global [12]. Their focus was on simply classifying queries, however, whereas we wanted to evaluate the geographic relevance of web pages.

2.2   Topic-Level Web Search

Topic-level web search has been an active research area recently. Its purpose is to provide knowledge on a certain topic, rather than to simply provide a large set of relevant pages.

Web QA is a typical example of topic-level web search system [9,10]. These systems answer a user's questions on a certain subject by gathering information from multiple pages.

Nie et al. recently used the term "web object" to refer to either a web page or a topic about which users can obtain information from web pages [7,8,6]. We use the term "web object" the same way here.

There has also been some work on the extraction of topic-level geographic knowledge from the web. Buyukkokten et al. observed a bias in the locations of sites linked to various newspaper sites [11]. Comparing the IP addresses of sites linked to the New York Times and the IP addresses of sites linked to the San Francisco Chronicle, they found that the sites linked to the New York Times were more widely distributed. This is an example of obtaining a topic's geographic relevance through web mining.

2.3   Geocoding

Geographic information retrieval systems must be able to map geographic names contained in text to coordinates. This is usually called *geocoding* [15] and it is done either by using addresses or by using place names in general. The former approach can provide high accuracy, but addresses are not always found in web pages. The latter approach is less accurate but is applicable to many pages without addresses.

Amitay et al. developed Web-a-Where, a system that maps web content to geographic space by using place names in general [16]. Lieberman and Sperling proposed STEWARD (Spatio-Textual Extraction on the Web Aiding Retrieval of Documents), an architecture for extracting addresses and place names by using a part-of-speech (POS) tagger and mapping web contents to geographic space using a feature vector model [17]. Sengar et al. worked on the disambiguation of place names appearing in search queries [18]. These research efforts, however, used rather ad-hoc methods to solve the problem of geocoding. In this paper we introduce a probabilistic model for geographic relevance and demonstrate its effectiveness experimentally.

2.4   GIS Modeling

Models similar to ours have also been used in GIS-related work. For example, membership functions are applied to model geographic relationships [19,20]. Their goal, however, was to model the ambiguity of geographic location, rather than to identify geographic regions relevant to pages or topics.

3   A Model for Geographic Relevance

In this section we describe a model of geographic relevance that web objects can have. We introduce the Gaussian mixture model as a model that fulfill a number of requirements for geographic relevance. We use this model throughout the rest of the paper.

3.1   The Use of Probability Density Functions

Geographic relevance refers to differences in the level of relevance among points in geographic space. For example, if a page is strongly relevant to a certain city, that is the pages' geographic relevance. Relevance to geographic objects with extensity (for example linear or areal objects) can be modeled as a summation of relevance to points.

One of the simplest models of geographic relevance is a function that assigns a level of relevance to each point in geographic space. We use probability density functions over geographic space as a model of geographic relevance. One merit of using a probabilistic model is that ambiguity in place names can be modeled as a noise in data.

3.2   Gaussian Mixture Models (GMM)

In this section we consider several requirements that a model for geographic relevance must satisfy.

One requirement is that the model must be able to incorporate two peaks in the level of relevance in geographic space. For example, "tropical rain forest" can be relevant to South America, Africa, and Southeast Asia. And the decrease in the level of relevance around peaks is not necessary isotropic.

Requirements for a model for geographic relevance:

- Level of relevance must be a function of geographic coordinates.
- Level of relevance may have multiple "peaks."
- Level of relevance decreases with increasing distance from a peak.
- Decrease in the level of relevance is not necessarily isotropic.

A Gaussian mixture model (GMM) satisfies these criteria[21]. It is a linear combination of multivariate Gaussian distributions, normalized so that the integral over the whole space R^d is equal to 1. Equation 1 is the definition of a Gaussian mixture model.

$$P(x) = \sum_{i=1}^{n} \alpha_i N(x; \mu_i, \Sigma_i), \qquad \sum_{i=1}^{n} \alpha_i = 1 \tag{1}$$

Equation 2 indicates a multivariate Gaussian distribution $N(x; \mu_i, \Sigma_i)$ with d dimensions. In the equation, $|\Sigma|$ indicates the determinant of Σ.

$$N(x; \mu_i, \Sigma_i) = \frac{\exp(-\frac{1}{2}(x - \mu_i)^T \Sigma_i^{-1}(x - \mu_i))}{(2\pi)^{\frac{d}{2}} |\Sigma_i|^{\frac{1}{2}}} \tag{2}$$

Since a Gaussian distribution is one of the simplest models in probability theory, it is widely used in applications. Here we model geographic space as two-dimensional and use a mixture of bivariate Gaussian distributions to represent geographic relevance.

4   Estimation of Geographic Relevance

In this section we describe a method for estimating the geographic relevance of web objects. Figure 1 illustrates the system flow.

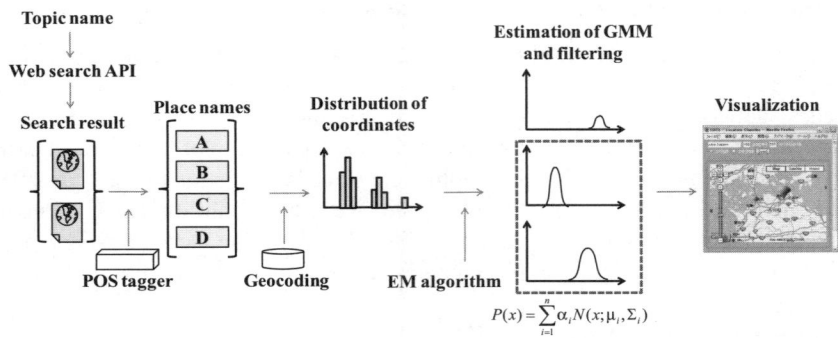

Fig. 1. System flow

4.1 Collection of Place Names from Web Pages

When a web page is the target of estimation, a robot retrieves the page and sends it to the part-of-speech tagging step. When the search target is a topic, the system uses a web search engine to collect pages relevant to the topic. A term that represents the topic is sent to the search engine as a query, and a list of URLs is retrieved.

Many search engines provide *snippets*, which are brief (about 100 characters) summaries of retrieved pages. These short texts are often more focused than the page itself and therefore more relevant to the search query. Since snippets can be obtained by accessing a search engine, their retrieval time is short. We therefore use snippets as the main source of data in our implementation.

In the next step, a conventional method of part-of-speech tagging is applied, and terms that were tagged as geographic objects are extracted.

4.2 Geocoding

In the geocoding step, place names appearing in the texts are mapped to geographic coordinates by using GIS (geographic information systems). When full addresses are used, there is little ambiguity and mapping is highly accurate. Many of the web pages that are relevant to a specific geographic area, however, do not contain addresses.

In this paper we focus on estimating geographic relevance from place names in general. This enables geographic relevance to be estimated from only a small amount of text. This, in turn, makes it possible to implement a real-time application based on a web search API.

When using place names in general, we encounter the problem of ambiguity. If a web page contains "Georgia", for example, it is not clear from the name alone whether it refers to a state in the U.S. or a country in the Caucasus. Place names are often more ambiguous than addresses, and an ambiguous place name does not identify a single location that a web page or topic is relevant to. Place names can, however, contribute to the process of estimating geographic relevance.

One advantage of using a probabilistic model is that the more place names we get, the better the estimate we can obtain. In this sense our method is suited for the geographic relevance of topics, since it is estimated using more documents for estimation, in comparison to the geographic relevance of web pages.

If we use a large set of documents, computation time would increase, especially that for the expectation-maximization (EM) algorithm. We reduce the computational load by sorting place names by their frequencies of appearance (text frequency) in the source data and using only the m place names with the highest frequencies.

When a place name w is not ambiguous, its text frequency $tf(w)$ is used as $\eta(w)$, the number that the element w in the sample space is observed. The value $\eta(w)$ is used for obtaining maximum likelihood through the EM algorithm. On the other hand, when a place name is ambiguous, $tf(w)$ is divided by the number of candidates $cand(w)$. Since $\eta(w)$ must be an integer, we must round it out as indicated in Equation 3. To reduce the effect of rounding, we multiply $\eta(w)$ by a weight factor γ beforehand. The reason for introducing the weight factor γ is to keep $\eta(w)$ greater than 1 even when $tf(w)$ is smaller than $cand(w)$. Since the number of observation $\eta(w)$ must be an integer, without the weight factor, cases where $tf(w) < cand(w)$ would all be rounded to 1, resulting in loss of information. In the equation below, w is a place name, $tf(w)$ is the frequency of w in texts, $cand(w)$ are the number of coordinates corresponding to w, and γ is a weight factor.

$$\eta(w) = \lceil \gamma \cdot tf(w)/cand(w) \rceil \tag{3}$$

4.3   EM Algorithm

As described in the previous section, our model for geographic relevance is expressed as a linear combination of multivariate Gaussian distributions, $N(x; \mu_i, \Sigma_i)$, with weights α_i. Parameters α_i, μ_i, and Σ_i are estimated using the EM algorithm [22].

A Gaussian mixture model is estimated using the EM algorithm. x_k represents an observation, m is the size of the sample, α_i is the weight for multivariate Gaussian distribution N_i, ϕ_i is the parameter vector for N_i consisting of μ_i and Σ_i. Φ is the set of all parameters in a GMM. In the following definition of the algorithm, a prime $'$ indicates that it is a value for the parameters in a prior step.

$$\alpha_i = \frac{1}{m} \sum_{k=1}^{m} \frac{\alpha_i' p_i(x_k | \phi_i')}{p(x_k | \Phi')} \tag{4}$$

$$\psi_{ik} = \frac{\alpha_i' p_i(x_k | \phi_i')}{p(x_k | \Phi')}, \quad \Psi_i = \sum_{k=1}^{m} \psi_{ik} \tag{5}$$

$$\mu_i = \frac{1}{\Psi_i} \sum_{k=1}^{m} \psi_{ik} x_k \tag{6}$$

$$\Sigma_i = \frac{1}{\Psi_i} \sum_{k=1}^{m} \psi_{ik} (x_k - \mu_i)(x_k - \mu_i)^T \tag{7}$$

Estimation of a Gaussian mixture model can be used as a clustering method in the following way. An observation x_k is assigned to a cluster $c_{i'}$ if and only if $N(x_k; \mu_{i'}, \Sigma_{i'})$ gives the highest probability among all Gaussians (Equation 8).

$$x_k \in c_{i'} \longleftrightarrow i' = \arg\max_i N(x_k; \mu_i, \Sigma_i) \tag{8}$$

4.4   Filtering

The content of a web page usually contains various place names, some of which are not used as place names. "China", for example, is a type of porcelain as well as a place name but also a type of porcelain, and "Washington" might refer to a person or a place.

To avoid inappropriate mapping caused by such ambiguities, we filter out clusters with small cardinality. In the filtering step, clusters are sorted by the number of elements they contain. Smaller clusters are filtered out, leaving only larger clusters while maintaining ratio r of the original sample as indicated in Equation 9, where X is the sample, K is a set of clusters before the filtering step, and c_i is a cluster in K. Clusters are sorted by their sizes so that $|c_i| \geq |c_{i+1}|$. After the filtering step, we obtain K', which is a set consisting of the larger clusters.

$$K' = \left\{ c_j \,\middle|\, \sum_{i=1}^{\xi} |c_i| \leq r|X| \wedge \sum_{i=1}^{\xi+1} |c_i| > r|X| \wedge j \leq \xi \right\} \tag{9}$$

We define a function $F(x)$ as a superposition of Gaussians $N(x; \mu_j, \Sigma_j)$ corresponding to clusters c_j in K':

$$F(x) = \sum_{c_j \in K'} \alpha_j N(x; \mu_j, \Sigma_j) \tag{10}$$

$F(x)$ is then normalized so that the integral over R^2 is 1. This is a requirement for a probability density function.

$$
\begin{aligned}
P(x) &= \frac{F(x)}{\int_{R^2} F(x)dx} \\
&= \sum_{c_j \in K'} \frac{\alpha_j}{\sum_{c_u \in K'} \alpha_u} N(x; \mu_j, \Sigma_j)
\end{aligned}
\tag{11}
$$

The probability density function $P(x)$ is the final output of the estimation.

5   Implementation

We have implemented a system that uses the method proposed in this paper. The system visualizes the geographic region relevant to a topic specified by a user and does this in real time. We call the system "LOCL" (for LOcation CLassifier). We built it using the AJAX architecture, with Perl on the server side and JavaScript on the client side. The system uses Google Maps API to show the base map [27].

5.1   Collecting Web Pages

The system uses Yahoo! Web Search API to collect web pages [26]. Compared to other web search APIs, Yahoo!'s API provides capability to set various parameters for searching, which makes it convenient in carrying out experiments. Texts can be obtained from the Web by collecting the snippets returned by a search engine, or by accessing pages themselves. While the first method is faster, the second is more accurate. We have implemented both methods in our system, and when submitting a query the user can choose which method to use. The user can also specify the number of the pages or snippets to be collected.

5.2   Extraction of Place Names

To extract place names, we apply MeCab, a part-of-speech tagger, to the search results [25] and extract words tagged "noun - proper noun - geographic object."

A series of numbers that matches the pattern of a postal code (zip code) is also extracted. This is possible because zip codes are expressed in a specific format. In case of Japan, it consists of three digits followed by a hyphen and four digits. The matched results may contain some noise (irrelevant numbers), but most of this kind of noise can be avoided by taking only the frequently occurring patterns.

Among the extracted place names, the m most frequent names are used in the following steps. We used $m = 100$ in our implementation. Place names that were ranked lower than 100 contained much noise and were often irrelevant to the area under focus.

5.3   Geocoding

In the first half of the geocoding step, the system maps names of administrative regions to coordinates by using the Digital National Information provided by the Geographical Survey Institute of Japan [24]. This data is freely available online. The administrative regions that we use are prefectures, cities, and villages.

If a region name is ambiguous, its frequency is divided by the number of candidate places as described in Subsection 4.2.

In the latter half of the geocoding step, place names that did not match administrative regions are mapped to coordinates by using a geocoding API provided by Yahoo! Inc. [26]. It contains famous buildings and landmarks. Like many other geocoding systems, it often returns multiple candidates for a single place name. Our system divides the original frequency by the number of candidate locations and lowers the frequency $\eta(x)$.

5.4   EM Algorithm

The system applies the EM algorithm to the set of coordinates and estimates a Gaussian mixture model. We used an original program written by Perl to perform the EM algorithm, and we used $\gamma = 6.0$ for the weight factor. Many of the place names with ambiguity are used to indicate two or three different locations. In these cases, multiplying $\eta(w)$ by 6 keeps the quotients to integers, and avoids the loss of information from rounding them. If we used a larger value for γ, it would increase the computation time. Using $\gamma = 6$ gave reasonable results while keeping the computation time small.

In the evaluation section, we will illustrate computation time for different values of γ. We have also set a maximum number in the iteration of the EM algorithm, to reduce the computation time.

5.5 Filtering

In the filtering step the system sorts clusters by their sizes and obtain the set of top ranked clusters that contains ratio r of the sample. In our implementation we used $r = 1/2$. The value was chosen as the simplest value between 0 and 1.

5.6 Visualization of Probability Density Function

Since a Gaussian distribution is represented by Expression 2, its equal probability density contour is a trajectory of Equation 12, where c is an arbitrary positive value.

$$(x - \mu_i)^T \Sigma_i^{-1}(x - \mu_i) = c \tag{12}$$

The trajectory draws an ellipse, having the major axis parallel to the eigenvector with the maximum eigenvalue and the minor axis parallel to eigenvector with the minimum eigenvalue.

In the map interface, each cluster is visualized using a marker and an ellipse. Figure 2 illustrates an example of visualization. The marker is placed on a location represented by the vector μ, and the ellipse is an equal-probability-density contour. The size of the marker represents the rank of the corresponding cluster when it is sorted by sample size. Figure 2 shows an example of visualizing search result for *udon* (Japanese noodle). Large clusters are formed in Kagawa and Osaka prefectures, which are both famous for noodle production and consumption.

By zooming in, the user can find relevant regions in a fine scale. Figure 3 shows a search result for regions within Kagawa prefecture that are relevant to *udon*.

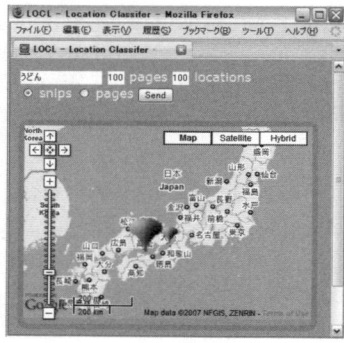

Fig. 2. Visualization interface

Fig. 3. Detailed search by zooming in

Fig. 4. Visualization of a geographic region relevant to a phrase

The system can also handle phrase queries. As illustrated in Figure 4, the phrase "I saw Mt. Fuji" is mapped to the region where Mt. Fuji can be seen.

6 Evaluation

We evaluated our method by estimating GMMs for various web pages and topics and checking whether the μ estimated for the largest cluster in a GMM is close to the intuitively correct location for the corresponding page or topic. We also evaluated the computation time.

6.1 Test Data Sets

For evaluating web pages, we listed around 40 URLs for each category. After estimation, we calculated the average accuracy and computation time. To make the experiment exhaustive we listed URLs selected from each of the prefectures in Japan.

Topics were also chosen from several categories, and we used Yahoo! Web Search API to obtain snippets for top 100 search results.

The categories used in the experiment are listed below.

For pages

- Tourist information sites (47 URLs)
- FM radio station sites (47 URLs)
- Shopping mall sites (47 URLs)

For topics

- Local cuisines (47 topics)
- Festivals (34 topics)
- Famous souvenirs (49 topics)

The list of topics for each category were taken from Wikipedia articles [28]. These articles describing the above mentioned categories give a long list of topics as examples, together with prefectures and cities that each topic is relevant to. Some of the examples taken from the lists are illustrated below.

Local cuisines
```
Potato cake            (Hokkaido prefecture)
Squid with miso paste (Aomori prefecture)
Stew in a stone pan    (Akita prefecture)
Aroid soup             (Iwate prefecture)
. . . . .
```

Festivals
```
Sapporo Snow Festival       (Hokkaido prefecture)
Nebuta Festival             (Aomori prefecture)
Lights-on-a-pole Festival (Akita prefecture)
Dressed-up Horse Festival (Iwate prefecture)
. . . . .
```

Experiments were carried out on a machine with a 3.20GHz Intel Xeon CPU, 8GB RAM, and CentOS release 5.

6.2   Amounts of Place Names and Coordinates Extracted

For evaluating estimation on pages, we used whole contents of pages. For evaluating estimation on topics, we used a set of snippets returned by web search API. Each web page contained 9.2 terms that were tagged as "noun - proper noun - geographic object", in average. For a snippet, 0.53 terms were tagged to this part of speech.

Table 1 illustrates the number of place names extracted from a page or a set of 100 snippets. It also shows the number of coordinates corresponding to the extracted place names. The numbers of place names differ between categories, and it is shown that each place name corresponds to a number of coordinates.

Table 1. Amounts of place names and coordinates

Category	Sample size	Avg. no. of place names	Avg. no. of coordinates
Pages 1	47	19.4	109.6
Pages 2	47	3.0	14.7
Pages 3	47	5.1	29.1
Topics 1	47	54.4	252.6
Topics 2	34	59.1	299.6
Topics 3	49	46.2	220.5

Pages 1: Tourist information sites, Pages 2: FM radio station sites, Pages 3: Shopping mall sites, Topic 1: Local cuisines, Topic 2: Festivals, Topic 3: Souvenirs (data on topics were taken from top 100 search results)

Table 2. Mapping accuracy

Category	Sample size	C1	C2	C3	C4
Pages 1	47	108.1	256.7	268.2	349.9
Pages 2	47	106.8	323.2	306.8	483.7
Pages 3	47	32.0	264.5	292.9	500.7
Topics 1	47	92.5	272.7	324.1	300.1
Topics 2	34	119.8	161.4	254.9	305.6
Topics 3	49	67.6	273.3	353.5	307.9

C1-C4: Clusters 1-4, Pages 1: Tourist information sites, Pages 2: FM radio station sites, Pages 3: Shopping mall sites, Topics 1: Local cuisines, Topics 2: Festivals, Topics 3: Souvenirs (distances in km)

6.3 Accuracy of Mapping

To evaluate accuracies of mapping on web pages and topics, we measured "error distances" between the estimated locations and the intuitively correct locations.

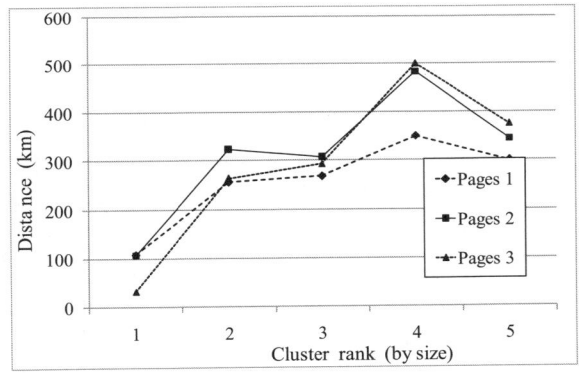

Fig. 5. Accuracy of mapping pages

For the intuitively correct locations ν we used the provincial government offices in corresponding administrative regions. Although it is not the spatial center of the region, usually it has the highest concentration of population in the region. We therefore considered it to be the representative location of the region.

Distances between ν and the centers of the jth largest clusters centers μ_j were measured, and averages were taken for each j.

The resulting average distances were (in decreasing order of cluster size), 82.3, 281.5, 289.3, and 444.8 km for pages and 90.4, 243.8, 317.1, and 304.5 km for topics.

Figures 5 and 6 illustrate the relations between error distance and cluster size. For all categories, the larger clusters were located closer to the intuitively correct locations. This indicates that removing smaller clusters is effective in estimating geographic

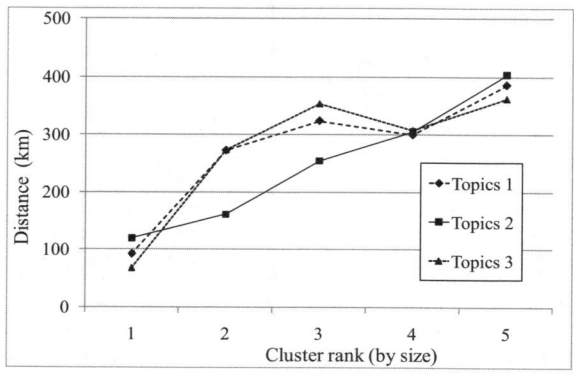

Fig. 6. Accuracy of mapping topics

relevance. The results for the largest clusters seem good when one considers that prefectures in Japan are often more than 100 km across.

6.4   Computation Time

The average computation time for each step is listed in Table 3.

Table 3. Computation time (seconds)

Category	Size	Search	Fetch	POS	Geocode	EM	Total
Pages 1	47	0.010	0.661	0.055	2.497	6.744	9.967
Pages 2	47	0.010	0.380	0.024	0.458	0.442	1.314
Pages 3	47	0.011	0.447	0.036	0.787	1.282	2.563
Average	47	0.010	0.496	0.038	1.247	2.823	4.615
Topics 1	47	8.105	0.001	0.418	5.929	65.187	79.640
Topics 2	34	12.598	0.001	0.444	5.847	158.041	176.931
Topics 3	49	2.517	0.001	0.422	4.303	87.808	95.051
Average	43.3	7.174	0.001	0.426	5.295	97.998	110.894

Search: web search, Fetch: page fetching, POS: part-of-speech tagging, Geocode: geocoding, EM: EM algorithm, Pages 1: Tourist information sites, Pages 2: FM radio station sites, Pages 3: Shopping mall sites, Topics 1: Local cuisines, Topics 2: Festivals, Topics 3: Souvenirs

The results show that most of the computation time is that for the EM algorithm. Figure 7 illustrates, for five different weight factors γ, how the average computation time (for five topics) increases with the number of place names used. With more place names and larger γ, the EM algorithm must classify more observations. Therefore the overall computation time increases. For this reason we have implemented our system in such a way that the user can select how many place names are to be used.

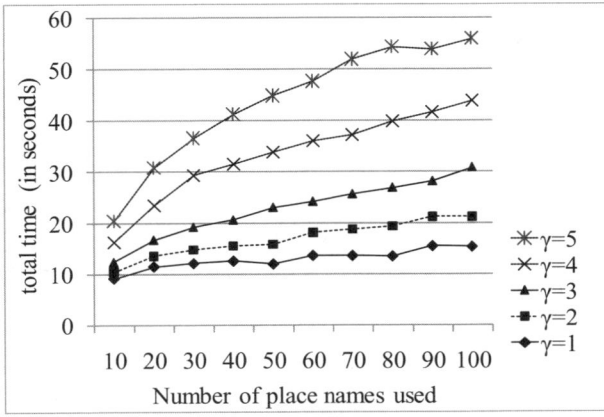

Fig. 7. Computation time vs. the number of place names used

7 Conclusion

In this paper we described a method for estimating the geographic relevance of web pages and topics and described an implementation of that method. Evaluation results showed that the center of the largest cluster was usually close to the intuitively correct location and that computation times were small enough for real-time application.

One problem with our present method is that it is vulnerable to a place name that is a homonym to a general noun, but is seldom used as a place name. In such case, the place name has a high text frequency but not considered as ambiguous, therefore weighed highly in the GMM estimation. We would like to cope with this problem by trying different filtering methods.

We are also extending our system to cover other countries, and we plan to implement a system that gathers relevant geographic regions for a large set of topics and stores them so that the system can provide topics to a user who submits a region as a query. Such a system could help a user find local specialties and would be useful for tourism.

Acknowledgment

This work was supported in part by a MEXT grant for "Development of Fundamental Software Technologies for Digital Archives," Software Technologies for Search and Integration across Heterogeneous-Media Archives (Project Leader: Katsumi Tanaka), a MEXT Grant-in-Aid for Scientific Research on Priority Areas: "Cyber Infrastructure for the Information-explosion Era," Planning Research: "Contents Fusion and Seamless Search for Information Explosion" (Project Leader: Katsumi Tanaka, A01-00-02, Grant#: 18049041), MEXT Grant-in-Aid for Scientific Research on Priority Areas: "Cyber Infrastructure for the Information-explosion Era," Planning Research: "Design and Development of Advanced IT Research Platform for Information" (Project Leader: Jun Adachi, Y00-01, Grant#: 18049073), and MEXT Grant-in-Aid for Young Scientists

(B) "Trust Decision Assistance for Web Utilization based on Information Integration." (Leader: Taro Tezuka, Grant Number: 18700086).

References

1. Raper, J.: Geographic relevance. Journal of Documentation 63(6), 836–852 (2007)
2. McCurley, K.S.: Geospatial mapping and navigation of the Web. In: Proceedings of the 10th International World Wide Web Conference, Hong Kong, China, pp. 221–229 (2001)
3. Gao, W., Lee, H.C., Miao, Y.: Geographically focused collaborative crawling. In: Proceedings of the 15th International World Wide Web Conference, Edinburgh, Scotland, pp. 287–296 (2006)
4. Zhou, Y., Xie, X., Wang, C., Gong, Y., Ma, W.Y.: Hybrid index structures for location-based web search. In: Proceedings of the 14th ACM International Conference on Information and Knowledge Management, Bremen, Germany, pp. 155–162 (2005)
5. Matsumoto, C., Ma, Q., Tanaka, K.: Web information retrieval based on the localness degree. In: Proceedings of the 13th International Conference on Database and Expert Systems Applications, Aix-en-Provence, France, pp. 172–181 (2004)
6. Chen, L., Zhang, L., Jing, F., Deng, K., Ma, W.Y.: Ranking web objects from multiple communities. In: Proceedings of the International Conference on Information and Knowledge Management, Arlington, Virginia, pp. 377–386 (2006)
7. Nie, Z., Ma, Y., Shi, S., Wen, J.R., Ma, W.Y.: Web object retrieval. In: Proceedings of the 16th International World Wide Web Conference, Banff, Canada, pp. 81–90 (2007)
8. Nie, Z., Wen, J.R., Ma, W.Y.: Object-level vertical search. In: Proceedings of the 3rd Biennial Conference on Innovative Data Systems Research, Asilomar, California, pp. 235–246 (2007)
9. Kwok, C., Etzioni, O., Weld, D.S.: Scaling question answering to the Web. In: Proceedings of the 10th International World Wide Web Conference, Hong Kong, pp. 150–161 (2001)
10. Radev, D., Fan, W., Qi, H., Wu, H., G.: Probabilistic question answering on the Web. Journal of the American Society for Information Science and Technology 56(6), 571–583 (2005)
11. Buyukkokten, O., Cho, J., Garcia-Molina, H., Gravano, L., Shivakumar, N.: Exploiting geographical location information of Web pages. In: Proceedings of the ACM SIGMOD Workshop on the Web and Databases, Philadelphia, Pennsylvania (1999)
12. Gravano, L., Hatzivassiloglou, V., Lichtenstein, R.: Categorizing web queries according to geographical locality. In: Proceedings of the 12th International Conference on Information and Knowledge Management, New Orleans, Lousiana, pp. 325–333 (2003)
13. Mei, Q., Liu, C., Su, H., Zhai, C.: A probabilistic approach to spatiotemporal theme pattern mining on weblogs. In: Proceedings of the 15th International World Wide Web Conference, Edinburgh, Scotland, pp. 533–542 (2006)
14. Tezuka, T., Kurashima, T., Tanaka, K.: Toward tighter integration of web search with a geographic information system. In: Proceedings of the 15th World Wide Web Conference, Edinburgh, Scotland, pp. 277–286 (2006)
15. Davis, C.A., Fonseca, F.T.: Assessing the certainty of locations produced by an address geocoding system. Geoinformatica 11(1), 103–129 (2007)
16. Amitay, E., Har'El, N., Sivan, R., Soffer, A.: Web-a-Where: geotagging web content. In: Proceedings of the 27th Annual International ACM SIGIR Conference on Research and Development in Information Retrieval, Sheffield, United Kingdom, pp. 273–280 (2004)
17. Lieberman, M.D., Sperling, J.: STEWARD: Architecture of a spatio-textual search engine. In: Proceedings of the 15th Annual ACM International Symposium on Advances in Geographic Information Systems, Seattle, Washington, Article No. 25 (2007)

18. Sengar, V., Joshi, T., Joy, J., Prakash, S., Toyama, K.: Robust location search from text queries. In: Proceedings of the 15th Annual ACM International Symposium on Advances in Geographic Information Systems, Seattle, Washington, Article No. 24 (2007)
19. Schneider, M.: Geographic data modeling: Fuzzy topological predicates, their properties, and their integration into query languages. In: Proceedings of the 9th ACM international symposium on advances in geographic information systems, Atlanta, Georgia, pp. 9–14 (2001)
20. Shi, W., Liu, K.: A fuzzy topology for computing the interior, boundary, and exterior of spatial objects quantitatively in GIS. Computers & Geosciences 33(7), 898–915 (2007)
21. Bishop, C.M.: Pattern recognition and machine learning. Springer, Heidelberg (2006)
22. Bilmes, J.A.: A gentle tutorial of the EM algorithm and its application to parameter estimation for Gaussian Mixture and Hidden Markov Models, Technical Report, University of Berkeley, ICSI-TR-97-021 (1997)
23. van Rijsbergen, C.J.: Information Retrieval - Second Edition. Butterworth & Co Publishers Ltd (1979)
24. Geographical Survey Institute of Japan, http://www.gsi.go.jp/ENGLISH/
25. MeCab, http://mecab.sourceforge.net/
26. Yahoo!, API, http://developer.yahoo.co.jp/
27. Google Maps, API, http://google.com/apis/maps/
28. Wikipedia, http://wikipedia.org/
29. Google Maps, http://maps.google.com/
30. Yahoo! Local Maps, http://map.yahoo.com/
31. Live Search Maps, http://maps.live.com/

Efficient Vessel Tracking with Accuracy Guarantees

Martin Redoutey[1], Eric Scotti[1], Christian Jensen[2],
Cyril Ray[1], and Christophe Claramunt[1]

[1] Naval Academy Research Institute, Brest Naval, France
{name}@ecole-navale.fr
[2] Department of Computer Science, Aalborg University, Denmark
csj@cs.aau.dk

Abstract. Safety and security are top concerns in maritime navigation, particularly as maritime traffic continues to grow and as crew sizes are reduced. The Automatic Identification System (AIS) plays a key role in regard to these concerns. This system, whose objective is in part to identify and locate vessels, transmits location-related information from vessels to ground stations that are part of a so-called Vessel Traffic Service (VTS), thus enabling these to track the movements of the vessels. This paper presents techniques that improve the existing AIS by offering better and guaranteed tracking accuracies at lower communication costs. The techniques employ movement predictions that are shared between vessels and the VTS. Empirical studies with a prototype implementation and real vessel data demonstrate that the techniques are capable of significantly improving the AIS.

Keywords: maritime navigation, tracking, trajectory prediction.

1 Introduction

Around eighty percent of all global interchange occurs via sea. One of the most successful systems used so far in maritime navigation is the Automatic Identification System (AIS), whose primary objectives are to identify and locate vessels at a distance. The AIS usually integrates a transceiver system as well as onboard GPS receivers and other navigational sensors such as gyrocompasses and rate of turn indicators. An onboard AIS transceiver operates in an autonomous and continuous mode, regularly broadcasting position reports according to the vessel's movement behavior. The reports are broadcast within a range of 35 miles to surrounding ships and Vessel Traffic Systems (i.e., maritime authorities) on the ground. The reports include the vessel's position, route, speed, and estimated arrival time at a port of call. The International Maritime Organization (IMO) has made the AIS a mandatory standard for the Safety of Life at Sea. As a result, passenger ships traveling internationally as well as cargo vessels with a gross tonnage above 300 tons are now equipped with AIS transponders.

In maritime areas with high densities of ships, the data volumes exchanged reach the inherent communication limits of the systems deployed, which entails losses of position reports that can adversely affect maritime safety. While solutions relying on increased numbers of communication bands have been proposed, the IMO is still asking for new approaches that can improve the performance of the AIS.

M. Bertolotto, C. Ray, and X. Li (Eds.): W2GIS 2008, LNCS 5373, pp. 140–151, 2008.
© Springer-Verlag Berlin Heidelberg 2008

This paper introduces tracking techniques based on shared predictions that aim to significantly reduce the amounts of position reports needed to accurately track vessels. The main principle is that a vessel and the surrounding infrastructure share a prediction of the vessel's near-future movement as well as a guaranteed accuracy. A Vessel Traffic Service then uses a vessel's prediction to determine the vessel's location, and the vessel transmits a new prediction as needed to ensure that the prediction never deviates from its actual location by more than the guaranteed accuracy. With good predictions, few position reports are needed.

The new techniques build on techniques previously developed for vehicles [1] and that follow recent advances in the development of logical models and physical structures for the efficient management of large volumes of location data (e.g., [4] [5] [6] [7] [8] [9] [12]). Other works in the field concern the introduction of novel algorithms to adjust incoming GPS data to route networks (e.g., [13] [14] [15]). Location data form trajectories that may be studied with purposes such as identifying emerging behaviors or reducing communication costs. The former includes the search for people displacements patterns at the local scale [10] [11] or even at the macro scale [16]. The latter is closely related to our study. Some solutions developed for vehicle tracking require a two-way communication between the server and the client. In contrast, our approach provides a solution to vessel tracking based on one-way client-to-server communication. The approach has been implemented and validated by a simulator that acts as a server in charge of vessels tracking. One of the key features of the algorithm presented is that it utilizes the best performing prediction technique among several alternatives, in accordance with the ship's behavior.

The reminder of the paper is organized as follows. Section 2 introduces the overall maritime tracking and prediction approach. Section 3 develops the principles of point-based and vector-based algorithms applied to maritime navigation and summarizes simulation results. Section 4 presents a decision tree along with optimization principles that identify the best prediction technique according to the ship's behavior. Finally, Section 5 draws conclusions.

2   Maritime Trajectory Prediction Principles

The AIS uses a VHF transceiver for automatically broadcasting position reports. The VHF signal is received by nearby ships and ground stations. The rate of transmission depends on the ship's current speed and maneuver, as illustrated in Table 1 [2].

The table gives the maximum time between successive updates as a function of the vessel's behavior, and also reports the resulting accuracies. The positions transmitted by the AIS are obtained using embedded GPS. The accuracy guaranteed by the AIS is the largest distance a given vessel can cover between two updates (IMO assumes an accuracy of 10 m for embedded GPS). It can be noted that there are no upper bounds guaranteed on the accuracy for the last two kinds of vessel behavior.

Let us assume that a server (e.g., a VTS or a vessel at sea) and a mobile object (e.g., vessel at sea/underway) are both able to predict the next position of the given vessel with a shared algorithm: this is a shared prediction system. The prediction of

Table 1. AIS update frequencies

Vessel behavior	Time between updates	Accuracy (m)
Anchored	3 min	10
Speed between 0 and14 knots	12 s	10-95
Speed at 0-14 knots and changing course	4 s	10-40
Speed at 14-23 knots	6 s	55-80
Speed at 14-23 knots and changing course	2 s	25-35
Speed over 23 knots	3 s	45-
Speed over 23 knots and changing course	2 s	35-

the next positions of the vessel is based on several steps as illustrated in Figure 1. When the server receives the position information from the vessel, it stores this information locally. Until the reception of the next update, it will use the information for predicting the vessel's position.

Using the same algorithm as the server, the tracked vessel regularly compares its GPS position with the predicted one. The vessel monitors the distance between the predicted position and the GPS position. When this distance exceeds an agreed upon threshold, an update is issued to the server. This is illustrated in Figure 2.

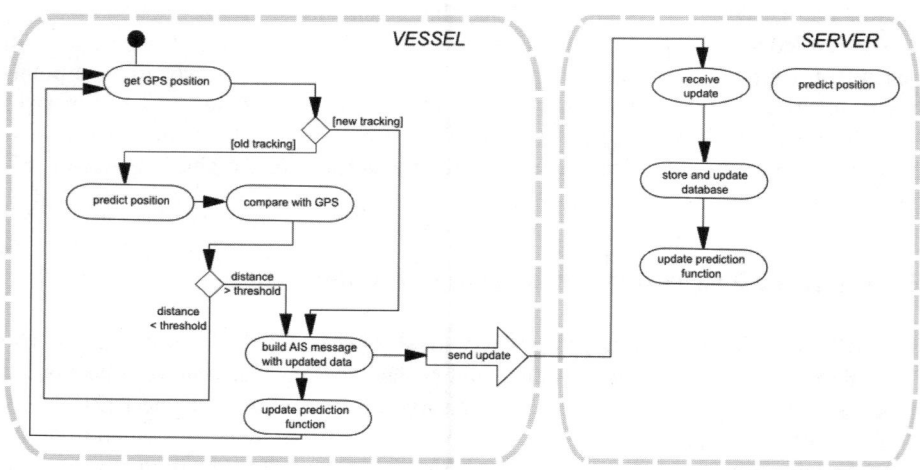

Fig. 1. Tracking principles

This approach to tracking is based on the assumption that client and server use the same prediction algorithm; however, the prediction algorithm can change in real-time as long as both sides work in concert and always use the same algorithm. The choice of which prediction algorithm to use should depend only on the data contained in an update. This data includes a GPS position, but it can also include heading, acceleration, and speed information. Given an update, the server and the vessel can then

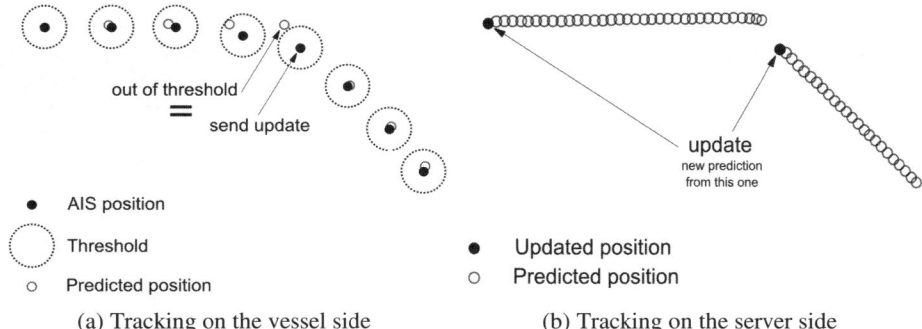

out of threshold

send update

● AIS position

⟨⟨⟨⟩⟩⟩ Threshold

○ Predicted position

● Updated position

○ Predicted position

update
new prediction
from this one

(a) Tracking on the vessel side   (b) Tracking on the server side

Fig. 2. Tracking mechanisms on the client and server sides

automatically switch to the same prediction algorithm (Figure 1). When the vessel receives the next GPS position, it makes a prediction from the last predicted or updated point using the newly selected algorithm. Similarly, the server uses its prediction algorithm continuously even if no GPS position has been transmitted by the vessel. In order to do so, the server uses (at least) the last reported position and the selected prediction algorithm.

3   Trajectory Tracking Strategies

Shared prediction-based tracking equipped with prediction algorithms that exploit the information broadcast by the AIS can significantly improve tracking accuracy while reducing communication. This section covers such prediction algorithms.

3.1   Point-Based Prediction

Point-based prediction algorithms predict that the current position of a vessel is the one contained in the most recent update. Thus, it is assumed that the vessel does not move. An update is sent by the vessel each time its GPS position differs from that of the last update by more than the current threshold. This prediction algorithm is expected to give good results for quite static objects. In a maritime context, this applies to moored or anchored vessels.

However, point-based prediction falls short when applied to objects with significant movement. In such settings, more sophisticated vector-based prediction is likely to perform better. This approach has to take advantage of additional location data contained in updates when forming predictions.

3.2   Vector-Based Prediction

With vector-based approach, predictions are linear functions of time. Considering the maritime domain and the AIS system, several specializations of the vector-based predictions used for vehicles are considered.

A first approach, called simply *vector-based*, computes a velocity vector using the last two reported positions (Figure 3): the predicted position is computed by a linear space-time function (assuming that the globe is locally plane) based on the positions in the two most recent updates (longitude, latitude, and time are required).

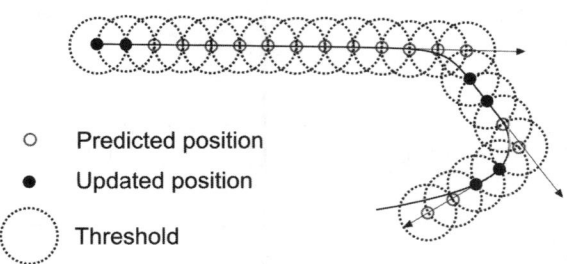

○   Predicted position

●   Updated position

◌   Threshold

Fig. 3. Vector-based prediction

A second approach uses the vessel's *heading*. This relies on the ability of the AIS to report a ship's heading and speed in every update. This data comes from the vessel's compass and loch systems, and it enables derivation of a velocity vector with a single update. The prediction is represented by a 4-tuple: [longitude, latitude, speed (knots), heading (degrees)] [17].

A third variation that exploits the data provided by the AIS uses the *course over ground* (COG) of a vessel. The two previous approaches ignore the drifting of a vessel as the heading does not consider sea currents. Using the course over ground solves that problem. Most GPS receivers aboard vessels can calculate the COG. When this information is available, it is transmitted by the AIS and is thus useable by the prediction algorithm.

3.3 Non-Linear Prediction

The vector-based approach can also be improved by taking into account *acceleration* information when a vessel exhibits a uniform acceleration (fast variations are difficult to identify with the AIS). This method can be efficient for ships getting underway or in the vicinity of harbors. Given two position updates and their speeds, the acceleration vector is easily determined. The prediction of the next position can be computed using the last COG or heading received and improved using acceleration information [17].

Tracking ships that turn can be done using the *rate of turn*. The next location of a turning ship can be derived by considering the heading discrepancy between two updates. Indeed, if this value has changed between two reports, it may be turning. But one need to know how fast the rotation of the ship is, and this cannot be determined with only two updates which can be far apart. That is why prediction algorithms using a ship's turning rate as included in AIS updates might be more efficient.

Figure 4 illustrates a turning ship. When an update occurs, the rate of turn has to be evaluated (not all AIS installations send such information). If it has a significant value, the predicted position is computed using the turning rate. A position is then given using a five-tuple: [longitude, latitude, speed (knots), COG or heading (degrees), rate of turn (°.min^{-1})] [17]. The ship is assumed to turn endlessly until a new update occurs, resulting in new values (Figure 4, location B). If the ship's rate of turn is null, the vessel must have taken a new route.

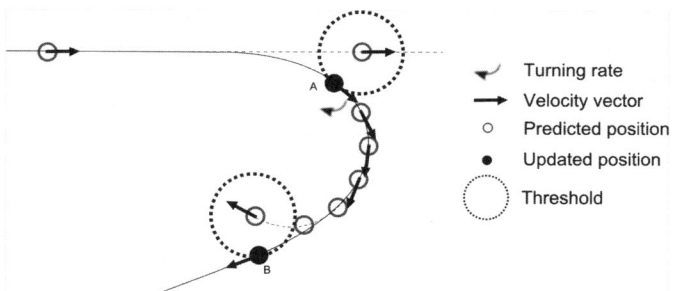

Fig. 4. Tracking of turning ships

3.4 Experimental Principles

The shared prediction approach and prediction algorithms have been prototyped. The simulation software relies on the *Poseidon platform* that gives access to a database of real-time AIS traffic data using Web-based services [3]. This database contains (1) vessels' static data such as the MMSI (international maritime identifier) and IMO identifiers, name, and ship type; (2) information about the journey such as destination, draught and estimated time of arrival; and (3) dynamic position reports that include time, longitude, latitude, heading, speed, course over ground, and rate of turn.

The simulation system allows the user to specify the relevant input parameters: the accuracy threshold to be used (based on the IMO assumption on intrinsic GPS accuracy, 10 m has been used in experiments), the specific AIS data to be used, and the prediction techniques to compare. The data used for the following experiments include speed, heading, course over ground, and rate of turn.

The time intervals between consecutive AIS reports exceed those of successive GPS reports. As the simulation system accesses broadcast AIS data located in the database, this means that the simulated vessel computes predicted position using AIS positions instead of using GPS data. Regarding a chosen threshold, this influences the accuracy and the detection mechanism to a smaller or larger degree. The results reported in the following section are nevertheless meaningful for comparing prediction techniques, although the results will be different when using GPS data. When embedding the algorithms in the AIS, they will obviously operate on GPS data.

3.5 Shared-Prediction Experiments

Empirical performance studies have been conducted on the point-based and vector-based prediction algorithms for different vessel behaviors: anchored, sailing straight, and changing speed and turning [17].

For anchored ships, the studies reveal that the algorithms perform similarly for small thresholds of 10 m (e.g., 10 m and an average update rate of 59% of that of the AIS). When the threshold is enlarged, point-based prediction performs better.

Regarding ships sailing straight, the studies show that point-based prediction is not appropriate, regardless of the speed. Heading-based and vector-based prediction perform well as far as drifting is minimized. When using a threshold of 57 m, they require only 4% and 7%, respectively, of the updates needed by the AIS. Cog-based prediction gives the best results and generates only 2% of the position updates needed by the AIS, assuming a 57 m threshold. It is worth to note that vector-based prediction is likely to give better results than heading-based prediction when updated positions are relatively close.

When the ship's speed is not constant, the studies show that cog-based and acceleration-based predictions are the best options. The factors that influence their relative performance are given by (1) the points selected for derivation of the acceleration, (2) how constant the acceleration is, (3) the type of trajectory of the ship (straight or curved), and (4) the threshold used.

Finally, when comparing the heading-based and cog-based predictions for turning ships, it has been found that even with large curves, cog-based prediction is best. This is because the AIS-based data influences the prediction mechanism by minimizing the number of locations taken into account, which is a problem for turning ships.

These studies show that the point-based and vector-based prediction algorithms take advantage of the availability of AIS data in many contexts. The policies are efficient within a given context and, overall, the findings can be summarized as follows:

- For ships anchored or docked, point-based prediction is the most efficient.
- When sailing straight, whatever the speed, vector-based predictions should be prioritized: first cog-based, then heading-based and finally vector-based technique.
- During accelerations and decelerations, the acceleration-based policy is efficient, but only when updated positions are available. Otherwise, the use of cog-based prediction should be retained.
- For ships changing their heading, no single type of prediction is a clear winner. Vector-based predictions such as heading-based and cog-based appear to be the most appropriate.

The empirical results might in some cases overestimate the "true" results. This is because relatively infrequent AIS data is used instead of data that is sampled frequently, e.g., each second. Overestimation occurs when the tracking threshold is exceeded before the time of the next AIS position. This affects experiments with small thresholds the most. For example, if the average time between two AIS positions is 10s, the curves may be overestimated by 10s in the worst case. If the average time between two updates is 500s, the error would be less than 2%. But if the average time between two updates is 30s, the error can reach 33%.

4   Combined Shared Prediction-Based Tracking

Based on findings of from the study covered in the previous section, we introduce a combined shared prediction-based approach that is capable of using different prediction algorithms in response to the movement behavior of a vessel. The objective of this approach is to determine in real time which prediction algorithm is best and then use that one, so that low cost (few updates) is achieved.

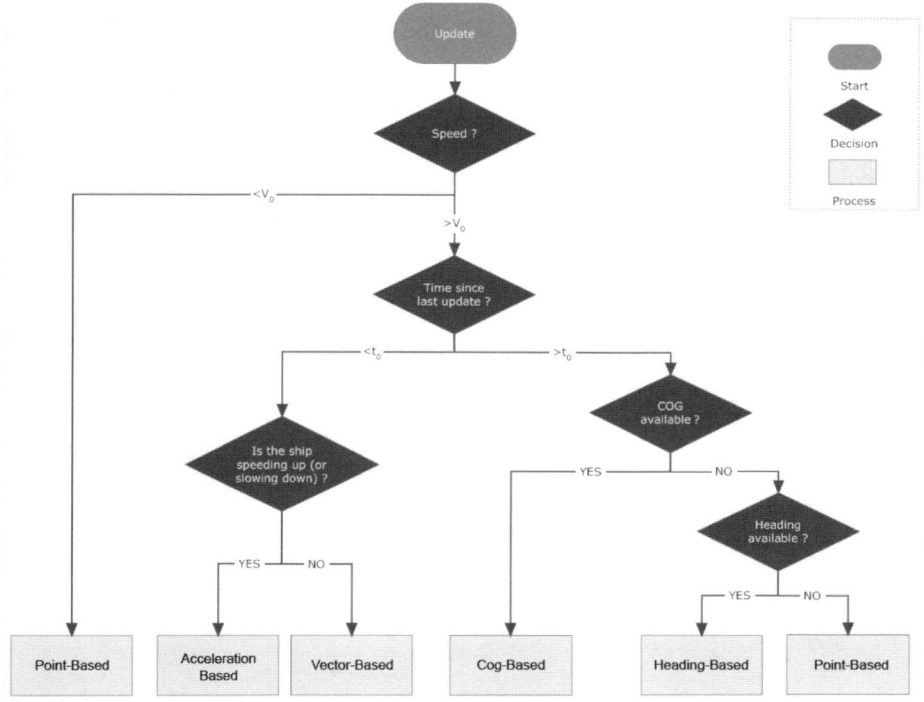

Fig. 5. Decision tree

This algorithm, illustrated in Figure 5, is based on the following variables:

- V_0: when a ship's velocity is below the value of this variable, the ship is considered to be moored or anchored. This velocity is set at 0.2 knots based on findings from the empirical studies.
- T_0: when the duration between the two most recent updates is below the value of this variable, either acceleration-based or vector-based prediction is activated. It is not straightforward to find an appropriate value for this variable, as the best value is likely to fluctuate even between AIS locations. Based on findings from the empirical studies, T_0 is set at 25 seconds.

The overall algorithm first evaluates the speed of the ship and then chooses either the vector-based or point-based approaches. Then it compares the time since the last update with T_0. When updates are close in time, we take that to indicate that the prediction algorithm used is inefficient. This means that the ship has accelerated, decelerated, or transmitted wrong data (e.g., the course over ground or heading). When two updates occur close in time, the overall algorithm will use acceleration-based or vector-based predictions. If not and if COG and heading values are available, a choice is made among cog-based, heading-based, and point-based prediction.

The combined algorithm has been compared to vector-based predictions using AIS data from the passenger ship *Enez Eussa 3*. This ship has been chosen as its trajectory embodies several mobility patterns. First, the ship is anchored, then it leaves Brest harbor accelerating and maneuvering; finally the ship sails straight and heads towards the Atlantic sea. Data from *Enez Eussa 3* incorporates each a range of possible errors, e.g., a speed of 0.1 knot when anchored, a course over ground that is not always available, and a heading that is at times unreliable.

Figure 6 illustrates the performance of the different predictions when using several thresholds. As expected, the best results (i.e., longest average time between updates) are obtained by the combined algorithm (i.e., using all prediction algorithms).

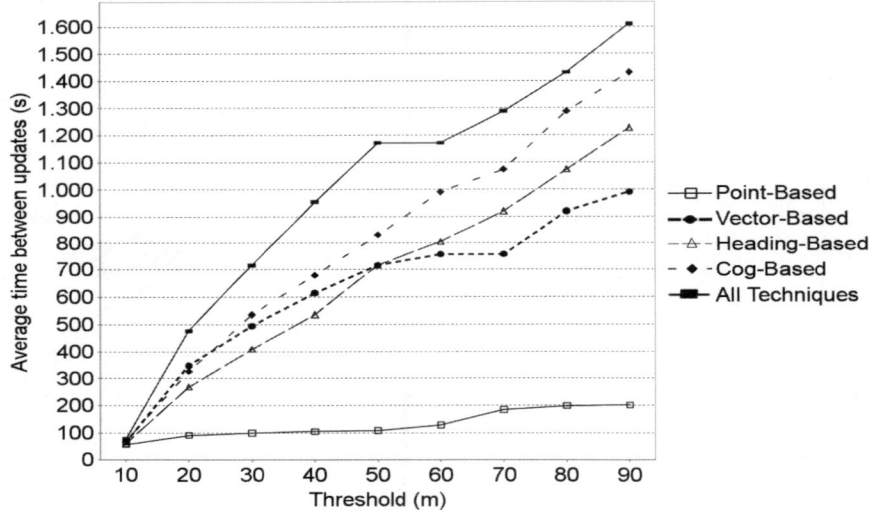

Fig. 6. Comparison of tracking algorithms for *ENEZ EUSSA 3* from 2007/05/18, 00:07:17 to 2007/05/18, 07:16:48

Table 2 shows the performance of the different tracking algorithms when applied to *Enez Eussa 3* (again from 2007/05/18 at 00:07:17 to 2007/05/18 at 07:16:48). The combined approach always yields the best results for threshold set to 10, 20, 40, and 80 m.

Table 2. Comparison of tracking algorithms (average durations between updates)

Threshold (m) Algorithm	10	20	40	80
Point-based	57s	91s	105s	198s
Vector-based	66s	348s	613s	920s
Heading-based	68s	268s	536s	1073s
Cog-based	69s	326s	678s	1288s
All combined	77s	477s	954s	1431s

The route followed by the *Enez Eussa 3* was analyzed in order to determine where the different prediction algorithms were used. Figure 7 represents the path followed by the ship when it left Brest harbor (fast-changing direction and low speed). The anchored position corresponds to the part identified by a square; here, point-based prediction was used. One can remark that during the first ship maneuvers, the combined algorithm mainly uses heading-based and vector-based predictions. Finally, while approaching the exit of the harbor, the algorithm started to favor cog-based prediction.

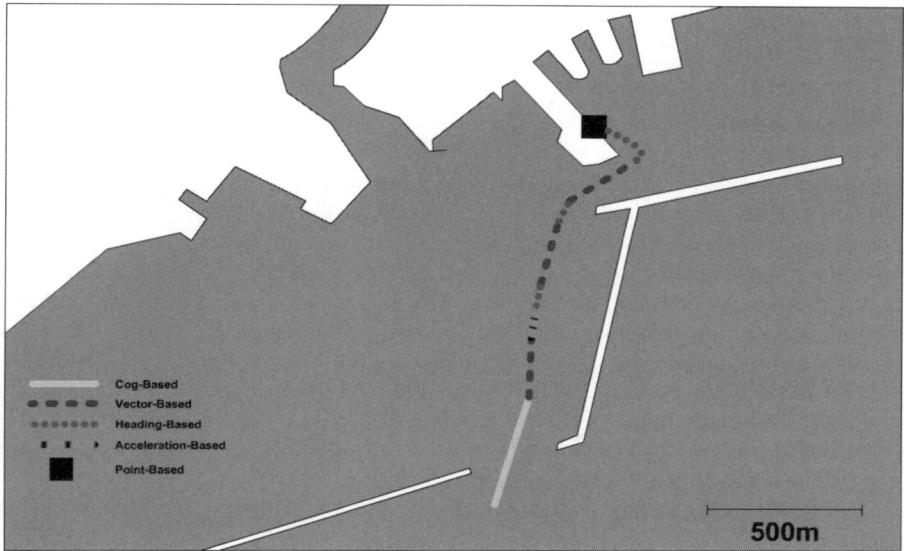

Fig. 7. Course of *Enez Eussa3* inside the harbor and prediction algorithms used

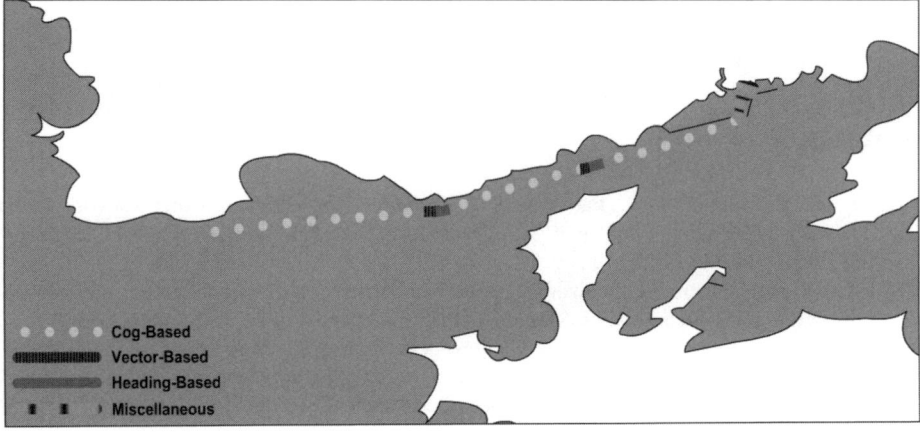

Fig. 8. Course of Enez Eussa 3 and policies used

Figure 8 illustrates the path of the ship and the techniques used from Brest harbor to the Atlantic sea. One can remark the uses of heading-based prediction for short periods of time. This comes from either a wrong heading being received or from drifting. In this navigation context (i.e., straight line and high speed), mainly cog-based tracking is used. When it is not used, this is due to the course over ground not being available.

Acceleration-based prediction is used only once. This is due to the fact that when considering an accelerating ship between two updates, the speed increases or decreases by more than 1 knot. This ship accelerates only slowly. Using a value below 1 knot may result in this type of prediction being overused, at the expense of vector-based prediction.

5 Conclusion

This paper introduces an efficient vessel tracking approach customized for the maritime environment, where vessels can be located using the international Automatic Identification System (AIS). The approach relies on several vessel position prediction algorithms that take advantage of the location data contained in AIS messages. By sharing predictions between the vessels and an on-ground Vessel Traffic Service, it is possible to save on the transmissions of position data. Empirical studies using real AIS data and simulated data show that point-based and vector-based predictions are efficient in different settings. A combined, context-aware algorithm is proposed that selects the best prediction algorithm according to the ship's movement behavior. The studies offer evidence that the amount of data transmitted by the AIS can be reduced very substantially, in many cases by more than 98%.

The combined algorithm can be extended, e.g., by considering the type and usual behavior of a ship and by using learning techniques. Trajectory patterns exhibited by common maritime routes and constraints derived from navigation rules may also be exploited for improving the tracking efficiency. Additional experimental studies are also in order. Current work is underway in the Brest harbor using the AIS system connected to mobile appliances and wireless communication (e.g., WIFI or ISM communications). The objective of these experiments is to evaluate update frequencies and appropriate thresholds in real contexts.

References

1. Civilis, A., Jensen, C.S., Pakalnis, S.: Techniques for Efficient Road-Network-Based Tracking of Moving Objects. IEEE Transactions on Knowledge and Data Engineering 17(5), 698–712 (2005)
2. International Maritime Organization, Guidelines for the onboard operational use of ship-borne Automatic Identification Systems (AIS), resolution A.917(22), 14 pages (2002)
3. Bertrand, F., Bouju, A., Claramunt, C., Devogele, T., Ray, C.: Web architectures for monitoring and visualizing mobile objects in maritime contexts. In: Proceedings of the 7th International Symposium on Web and Wireless Geographical Information Systems, pp. 94–105 (2007)

4. Saltenis, S., Jensen, C.S., Leutenegger, S.T., Lopez, M.A.: Indexing the positions of continuously moving objects. In: Proceedings of ACM SIGMOD International Conference on Management of Data, pp. 331–342 (2000)
5. Kollios, G., Gunopulos, D., Tsotras, V.J.: On indexing mobile objects. In: Proceedings of the 18th ACM Symposium on Principles of Database Systems, pp. 261–272 (1999)
6. Sistla, A.P., Wolfson, O., Chamberlain, S., Dao, S.: Modeling and querying moving objects. In: Proceedings of the International Conference on Data Engineering, pp. 422–432 (1997)
7. Ding, Z., Güting, R.H.: Managing moving objects on dynamic transportation networks. In: Proceedings of the International Conference on Scientific and Statistical Databases, pp. 287–296 (2004)
8. Güting, R.H., Böhlen, M.H., Erwig, M., Jensen, C.S., Lorentzos, N.A., Schneider, M., Vazirgiannis, M.: A foundation for representing and quering moving objects. ACM Trans. Database Syst. 25(1), 1–42 (2000)
9. Jensen, C.S., Lin, D., Ooi, B.C.: Query and update efficient B+-tree based indexing of moving objects. In: Proceedings of International conference on Very Large Data Bases, pp. 768–779 (2004)
10. Ashbrook, D., Starner, T.: Using GPS to learn significant locations and predict movement across multiple users. Personal and Ubiquitous Computing 7(5), 275–286 (2003)
11. Liaoa, L., Patterson, D.J., Foxa, D., Kautz, H.: Learning and inferring transportation routines. Artificial Intelligence 171(5-6), 311–331 (2007)
12. Lazaridis, I., Kriengkrai, P., Mehrotra, S.: Dynamic Queries over Mobile Objects. In: Proceedings of the 8th International Conference on Extending Database Technology, pp. 269–286 (2002)
13. Bernstein, D., Kornhauser, A.: An Introduction to map matching for personal navigation assistants. New Jersey TIDE Center, 17 pages (1996)
14. Yin, H., Wolfson, O.: Weight-based map matching method in moving objects databases. In: Proceedings of the International Conference on Statistical and Scientific Database Management, pp. 437–438 (2004)
15. Taylor, G., Blewitt, G.: Intelligent positioning: GIS-GPS Unification. 181 pages. Wiley, Chichester (2006)
16. Thériault, M., Claramunt, C., Seguin, A.-M., Villeneuve, P.: Temporal GIS and statistical modelling of personal lifelines. In: Ottawa, D.R., van Oosterom, P. (eds.) 9th Spatial Data Handling symposium, 9-12 July, pp. 433–450. Springer, Heidelberg (2002)
17. Redoutey, M., Scotty, E.: Tracking vessels with low cost and guaranteed high accuracy, Technical Report, Aalborg University and Naval Academy Research Institute, 52 pages (2008)

Spatial Factors Affecting User's Perception in Map Simplification: An Empirical Analysis

Vincenzo Del Fatto, Luca Paolino, Monica Sebillo,
Giuliana Vitiello, and Genoveffa Tortora

Dipartimento di Matematica e Informatica, Università di Salerno
84084 Fisciano (SA), Italy
{vdelfat,lpaolino,msebillo,gvitiello,tortora}@unisa.it

Abstract. In this paper, we describe an empirical study we conducted on the application of a simplification algorithm, meant to understand which factors affect the human's perception of map changes. In the study, three main factors have been taken into account, namely Number of Polygons, Number of Vertices and Screen Resolution. An analysis of variance (ANOVA) test has been applied in order to compute such evaluations. As a result, number of vertices and screen resolution turn out to be effective factors influencing the human's perception while number of polygons as well as interaction among the factors do not have any impact on the measure.

Keywords: Ramer-Douglas-Peucker Algorithm, Human Factors, Cognitive Aspects, Maps, Controlled Experiment, ANOVA Test.

1 Introduction

In the area of geographic information management, *simplification* of lines is a *generalization process* that aims to eliminate the "unwanted" map details.

That is to say, a simplification algorithm is a method to select a subset of points which best represents the geometric properties of a polyline while eliminating the remaining points.

In spite of the apparent straightforwardness of the concept of map simplification, this process may become very complex and is presently addressed as a solution to several interesting research issues in the field of spatial data management. As a matter of fact, simplification commonly yields:

- Quick Map Transmission: Simplified maps may be quickly transmitted over the network with respect to original maps because they are slighter in terms of number of vertices.
- Plotting time reduction– If the plotting process is slow, it may cause a bottleneck effect. After simplification, the number of vertices is reduced, therefore decreasing the plotting time.
- Storage space reduction – Coordinate pairs take up large amounts of space in GISs. The coordinates or data can be significantly reduced by simplification, in turn decreasing the amount of storage space needed and the cost of storing it.

M. Bertolotto, C. Ray, and X. Li (Eds.): W2GIS 2008, LNCS 5373, pp. 152–163, 2008.
© Springer-Verlag Berlin Heidelberg 2008

- Quick Data Processing: Simplification can greatly reduce the time needed for vector processing a data set. It can also speed up many types of symbol-generation techniques.

The problem focuses on choosing a threshold indicating what simplification level we consider acceptable, that is to say the rate of how many details we are ready to sacrifice in order to obtain a faster plotting, a faster transmission, or a reduced storage space for our geometric data. Generally, choosing such a threshold is a complex problem and users are usually let to decide on a case-by-case basis, and very little human guidance or automated help is generally available for this.

In this paper, we carried out an empirical study aimed at finding out which factors affect the choice of the threshold when the final user's aim is just to see the original map and no post-elaboration has to be done. That is to say, we performed an empirical study with the goal to understand which factors influence the map reduction threshold by measuring the subject's perception of changes of maps. Such a comprehension is fundamental to perform successive regression analysis in order to extract a general rule for calculating this threshold.

One of the most important tasks we faced during the experiment design phase was producing the maps which could be shown to the subjects in order to highlight which factors mostly affect their perception. Once determined such factors (number of vertices (NV), number of polygons (NP) and screen resolution (SR)) we chose some maps which satisfy any combination of the factor levels. Therefore, for each map we generated some simplified versions at fixed rates by means of a web application which computes the Douglas-Peucker (RDP) algorithm [4,16]4.

The remainder of this paper is divided as follows. In Section 2, we present some preliminaries. Then, Section 3 introduces the experiment design, we present all the choices we have made and the tests we have performed to prove our assertions. Section 4 presents the result of the test we performed along with their interpretations. A discussion on the achieved results concludes the paper in Section 5.

2 Preliminaries

Map Simplification
Contour lines are the model generally used for geographic features representation of digital maps in Geographical Information Systems (GIS). How to simplify map contour lines is a very popular research issue in the field of Map Generalization [2,12]. In particular, the need to provide on-the-fly the user with the same map at different scales makes the automated map simplification a relevant *Map Generalization* technique. Many methods have been developed to perform map simplification [9,5], such as, among the most cited, the Visvalingam-Whyatt algorithm [22] and its extension [24]. Nevertheless, the method most commonly adopted within GIS and Cartographic tools and applications is certainly the Ramer-Douglas-Peucker (RDP) algorithm [4,16] and its improvements [7,26,17]. This algorithm uses a threshold in order to remove line vertices from the map. Such a characteristic makes the simplification process by means of the RDP algorithm difficult to tune [5], because no linear relationship exists between the threshold to choose and needed map scale [3]. To the best of our knowledge, just one study exists in literature to automate this process, namely, Zhao *et al.* [26] use

the Topfer's Radical Law [21] to modify the RDP algorithm in order to automatically get the threshold. In other studies the threshold is selected in empirical way. In [10] a threshold of 25 is chosen for data in 1:500,000 scale, while a threshold of 35 for data in 1:1,000,000 scale. Other studies [23,15] employ similar ranges of threshold. With regards to commercial applications two available solutions exist. ArcGIS™ provides the *Pointremove* and *Bendsimplify* operators which are both based on the RDP algorithm. The former removes vertices quite effectively by a line thinning/compression method. It achieves satisfactory results but produces angularity along lines that is not very pleasing from an aesthetic point of view. In addition, in some instances, cartographic manual editing is required. The latter operator is designed to preserve cartographic quality. It removes small bends along the lines and maintains the smoothness [8].

State of the Art
As for research relating maps with the study of cognitive aspects, much work has been done especially in map design [13, 11] and in usability studies based on cartographic human factor [19]. However, concerning generalization and simplification, the literature seems to be limited to few studies in cognitive aspects of map scale [14]. These studies concern with general aspects of this topic, like what people means for basic conceptual structure of scale, size, resolution, and detail; difficulty of comprehending scale translations on maps; definition of psychological scale classes; definition of the relationship between scale and mental imagery; etc, but do not take into account how users perceive map simplification.

3 Design of the Experiment

In order to understand which factors affect the user's perception of map variation when maps are simplified by means of the Douglas-Peucker algorithm, we have performed a controlled experiment in which the influences of some key factors have been measured. In particular, in this experiment we focus our attention on three important features of the visualized map, namely the number of vertices (NV), the number of polygons (NP) and the screen resolution (SR). Some other factors might influence variation perception, such as roughness or map size. As an example, map size may affect the user perception because, it is evident that, the smaller the map, the less details the that the subject is able to perceive. However, we decided not to include the influence of this factor in our initial conjectures, and, in order to avoid that it biased the results we gathered, we decided to present maps according to a predefined size, namely 26cm x 15cm. Besides verifying whether the considered properties affect the user's perception of map variation, we also wished to measure the degree of influence of each factor in order to find out whether some general rules exist for optimizing the reduction factor on the basis of the map properties. In the following we formally describe the experiment by highlighting the most important choices we had to face.

Independent Variables. In our experiment, the independent variables are *NV*, *NP* and *RES*, while the dependent one is the Significant Simplification Rate (SSR), which is defined as the RDP simplification rate that we observe when the subject starts to see evident changes in a reduced map with respect to the original map.

Number of Vertices - NV. The number of vertices is the first independent variable we took into account in our experiment. For the sake of clarity, we call vertex a point which lies on either the start or the end of a map segment. Usually, this value may be represented by an integer variable in the 0-infinite range. However, for our purpose we decided to segment the range of possible values into four categories, namely: (1) from 1000 to 2000, (2) from7000 to 8000, (3) from 46000 to 47000 and (4) from 100000 to 120000. Figure 1 shows an example of a map used in the experiment with a 40% simplification map.

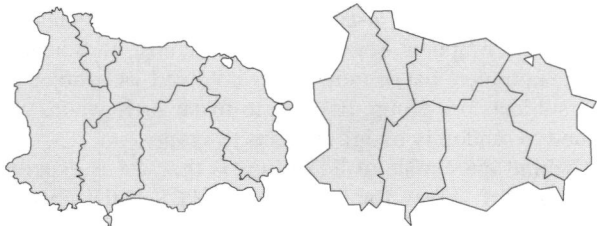

Fig. 1. An example map represented through two different sets of vertices

Number of Polygons – NP. The number of polygons is the second variable we investigated. It seems plausible that this factor affects the user's perception because the higher the number of objects visualized, the more detailed the map's perception. Similarly to the number of vertices, we decided to categorize the NP variable with the following three different levels: (1) less than, or equal to, 5 polygons, (2) from 6 to 10 polygons and (3) from 11 to 30 polygons.

Screen Resolution – RES. The third variable corresponds to the size of the screen where maps are visualized. By reducing the number of pixels used for visualizing the map, some vertices may overlap, thus affecting user's perception. However, apparent overlapping can be resolved by increasing the resolution. This in mind, we considered 2 screen resolutions (levels), namely 800x600 and 1280x768, the most commonly used.

Dependent variable. Usually during a map simplification (reduction of the number of points required to draw the map) process, the user's perception of changes may vary from very little perception to a point when the simplification becomes evident. Generally, we may identify the following five levels of perception of changes:

1. *No simplification detected*, in this case the subject does not recognize any difference between the original map and the simplified map.
2. *Some minor simplifications detected* which do not affect the appearance of the map. In this case, some differences are perceived but they are not relevant. As an example, simple linearization of raw lines or smaller contour lines due to less overlapped points.
3. *Some evident simplifications detected* which do not alter the meaning of the map. For instance, when rough lines get fairly straight.

4. *Some substantial simplifications detected* which affect the information the map conveys. The map is still recognizable but some changes may hide relevant details. As examples, some peninsulas may disappear; boundaries may deeply alter their position, and so on.

5. The map can no longer be recognized.

Based on the above classification, we defined the *Significant Simplification Rate (SSR)* as the simplification rate which makes the user's perception move from level 2 to level 3.

Participants. The participants we used in this research were students following the degree programme of Computer Science at the faculty of Science, University of Salerno (Italy). The complete list of individuals involved 144 subjects divided into 24 groups, namely 6 subjects per group. In order to make such groups as independent as possible we decided to randomly assign subjects to groups.

The rationale behind the choice of 24 groups is that 24 is exactly the number of different combinations of values for *NP*, *NV* and *RES*. All subjects were asked to complete a background information sheet in order to collect both personal data and data about their experiences with computer technology and, possibly, with GIS, also in terms of the exams they had already passed and the marks they had obtained.

We were interested in understanding how people who generally have marginal experience with maps perceive different levels of simplification. Consequently, we avoided subjects who work with maps or had attended some GIS courses. Actually, the individuals we selected knew the most popular map applications, such as some web-based path finders, Google Map, etc. As for their age, they ranged between 20 and 30 years old.

Apparatus. As for the apparatus, we exploited three software applications, namely, *SYSTAT*™ [20], *MapShaper* [1] and *ESRI ArcView*™ [6]. They were run on a Windows XP© Professional platform mounted on a computer based on a Pentium Centrino™ processor with 1G RAM, a 72 Gb HD at 7200 rpm and a 15.4'' multi resolution display. We used MapShaper for producing the reduced maps. It is a free online editor for Polygon and Polyline Shapefiles. It has a Flash interface that runs in an ordinary web browser. Mapshaper supports the Douglas-Peucker (RDP) line simplification algorithm. Finally, we used the ESRI ArcView 3.2 to produce the source maps we used in the experiment and to express them also in terms of the combination of the factor levels for, respectively, the number of polygons (*NP*) the number of vertices (*NV*) and the screen resolution (*SR*).

Task. Every group was associated with a map which satisfies one of the 24 combinations of the independent variable levels. This map and a simplified version of it were successively shown to each subject in the group, for ten second intervals. Then, according to the scale concerning the perception of changes, we asked the subjects for a value between 1 and 5. This step was repeated, increasing the simplification rate, until either subjects reached the maximum simplification rate or rated 5 their perception of changes. For each subject we reported the value of SSR, namely the simplification rate registered when they detected some evident simplifications, which did not alter the meaning of the map (i.e., level 3 of perception).

Test. Once results have been collected we applied a three way ANOVA (Analysis of Variance). The theory and methodology of ANOVA was mainly developed by Fisher during the 1920s [18]. ANOVA is essentially a method for analyzing one, two, or more quantitative or qualitative variables on one quantitative response. ANOVA is useful in a range of disciplines when it is suspected that one or more factors might affect a response. ANOVA is essentially a method of analyzing the variance of a response, dividing it into the various components corresponding to the sources of variation, which can be identified. We are interested in studying two effects - the main effect determined by each factor separately and the possible interactions among the factors. The main effect is defined to be the change of the evaluated factor (the dependent variable) while changing the level of one of the independent variables. In some experiments, we may find that the difference in the response between the levels of one factor is not the same at all levels of the other factors. When this happens, some interaction occurs between these factors. Moreover, the ANOVA method allows us to determine whether a change in the responses is due to a change in a factor or due to a wrong sample selection.

Research Hypotheses. Before starting the operative phase, we formally claim the null hypotheses we wished to reject in our experiment.

Table 1. Experiment hypothesis

H1. There is no difference in the SSR means of the different levels of factor NV
H2. There is no difference in the SSR means of the different levels of factor NP
H3. There is no difference in the SRR means of the different levels of factor RES
H4. There is no interaction between NP and NV on SSR.
H5. There is no interaction between NV and RES on SSR.
H6. There is no interaction between RES and NP on SSR.
H7. The two-way NV*NP interaction is the same at every level of RES

As listed in Table 1, the first three hypotheses are concerned with the influence that each factor may have on the SSR measure, separately. The next three hypotheses are concerned with the pair-wise interaction between independent factors and its influence on SSR. Finally the last hypothesis involves the study of the interaction among all three independent variables and its influence on SSR.

Assumptions. The ANOVA test is based on the following assumptions about the input data: (1) All observations are mutually independent; (2) All sample populations are normally distributed; (3) All sample populations have equal variance. The first assumption should be true on the basis of how we choose individuals and how we assigned them to groups.

The *Shapiro-Wilk's test* (W) [18] is used in testing for normality. It is the preferred test of normality because of its good properties as compared to a wide range of alternative tests. It is a standard test for normality used when the sample size is between 3 and 5000. The W-value given by this test is an indication of how good the fit is to a normal variable, namely the closer the W-value is to 1, the better the fit is. In our case, high values of W are common to all the groups we took into account; therefore we should accept the hypothesis of normality for each group. Table 2 shows the W values obtained after having applied the *Shapiro-Wilk Test* to each group.

Table 2. Results of the Shapiro-Wilk Test

NP	NV	W (800x600)	W (1024x768)
<5	1k-2k	0.7	1.0
6-10	1k-2k	0.7	0.8
11-30	1k-2k	1.0	0.7
<5	7k-8k	0.9	0.8
6-10	7k-8k	0.9	0.8
11-30	7k-8k	1.0	0.9
<5	46k-47k	0.8	0.9
6-10	46k-47k	0.7	0.8
11-30	46k-47k	1.0	0.8
<5	110k-120k	0.8	1.0
6-10	110k-120k	0.9	1.0
11-30	110k-120k	0.5	0.9

Next, we see the *Levene's Test* [18] for equality of variances. This tells us whether we have met our third assumption, namely whether the groups have approximately equal variance with respect to the *SSR*.

If the *Levene's Test* is significant (the value under p is less than 0.05), the variances are significantly different. If it is not significant (p is greater than 0.05), the variances are not significantly different; that is, the variances are approximately equal. In our case, we see that the significance is 0.6, which is greater than 0.05. Thus, we can assume that the variances are approximately the same.

Since the controls we made confirm our expectations, we may claim that assumptions have been respected and that data and groups are proper for the analysis we want to perform.

Results. The ANOVA Table 3[1] shows, for each factor, the number of degree of freedom (DF), the sum of squares (SS), the mean of the squares (Mean Square), the value of the statistical (F-Ratio), its significance level (Sig. Level) and the F critical value: if the Sig. Level is smaller than 0.05, then the factor effect is statistically significant at the level of confidence of 95%. The significant factors and their significance levels are highlighted in boldface in the tables. This kind of tabular representation is customarily used to set out the results of ANOVA calculations.

---

[1] In this test we firstly fixed the acceptable error (5%) which together with the Df implies the F-critical. If the calculated F-ratio is greater than the F-critical means that the test is significative and the null hypothesis should be rejected. The p-value specifies the probability that the F-ratio is greater of the F-critical. Generally, a p-value smaller than 0.05 tell us that this probability is greater than 95%. Type III SS and Mean Squares are intermediate values required to calculate the F-Ratio.

Table 3. A summarizing table showing the results of the ANOVA test. Values of p less then 0.0.5 indicate that the hypothesis should be rejected.

Hyp.	Factor	Type III SS	Df	M.S.	F-Ratio	P-Value	F critical
H1	**NV**	**1327.1**	**3**	**442.4**	**65.8**	**0.0**	**2.68**
H2	NP	15.2	2	7.6	1.1	0.3	3.07
H3	**RES**	**70.8**	**1**	**70.8**	**10.5**	**0.0**	**3.92**
H4	RES*NV	5.2	3	1.7	0.3	0.9	2.68
H5	RES*NP	6.7	2	3.4	0.5	0.6	3.07
H6	NV*NP	44.6	6	7.4	1.1	0.4	2.18
H7	RES*NV*NP	16.3	6	2.7	0.4	0.9	2.18

The interpretation of the ANOVA test is backward, that is we have to firstly verify whether interactions exist between both any pair of the factors and all the factors. In this case, we should take into account just the interaction and we could not claim any conclusion regarding the main effects. As a matter of fact, when interactions happen, the effect of a factor is conditioned by the level of the other one. By looking at Table 3, we notice that the Hypothesis H4, H5 and H6 which verify the interaction between each pair of factors and H7 which verifies the interaction among all the factors at the same time should be accepted because the F-Ratio are lower with respect to the corresponding critical F. It means that no factor is influenced by the change of level of any other factor and that each factor provides the perception with its effect alone.

For this reason, from now we focus our attention on verifying the main effect due to the single factors. As for the Hypotheses H1, which verifies whether the NV affects *SSR*, results prove that it can be rejected. As a matter of fact, the row corresponding to this hypotheses tells us that the F-Ratio is greater than the critical value $F(3, 120)$ in correspondence of a p significance value of 0.0. In this case, the factor effect is found to be significant, it implies that the means values differ more than it would be expected by chance alone. It is also evident by Table 4 and Figures 2(a) and (b) that by increasing the number of points and keeping constant the number of polygons, the simplification rate increases with both 800x600 resolution and 1024x768 resolution. In particular, the *SSR* constantly increases until reaching a critical value over which it smoothly gets flat. Actually, the curve of the perception seems to be approximated by a logarithmic function.

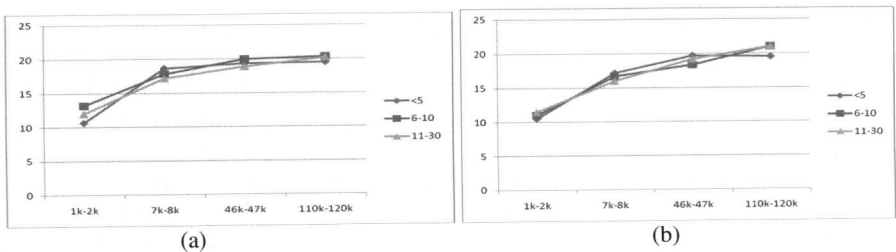

Fig. 2. (a) The RDP rate reduction lossless of significant details at 800x600 RES level, (b) The RDP rate reduction lossless of significant details at 1024x768 *RES* level

Table 4. Statistical about NV factor

Factor	Level	LSM	SErr	N
NV	1K-2K	11.5	0.4	36
NV	7K-8K	17.2	0.4	36
NV	46K-47K	18.7	0.4	36
NV	110K-120K	19.0	0.4	36

- 1K and 2K may be approximately reduced around at 65% (11.5 RDP rate) without loss of significant details,
- 7K and 8K may be simplified around at 72% (17.2 RDP rate),
- 46K and 47K may be simplified around at 72% (18.7 RDP rate) and
- 110K and 120K may be simplified around at 76% (19.0 RDP rate).

By taking into account such values and their meanings we can observe that any map we are managing, which has more than 1K vertices, may be approximated by a reduced one having less than 65% of the original vertices. As an example, a map having 1000 vertices may be reduced through the RDP algorithm to a map having 350 vertices without loss of significant details. As the number of vertices increases this percentage slightly augments as shown in the previous figures. As for Hypotheses H2, results are summarized in Table 3 and 4, Figures 3(a) and (b).

Table 5. Statistical about the *Number of Polygons* factor

Factor	Level	LSM	SErr	N
NP	A < 5	16.6	0.4	48
NP	B 6-10	17.0	0.4	48
NP	C 11-30	16.2	0.4	48

Table 1 highlights that we cannot reject this hypothesis and this factor does not affect the perception of the subjects. This claim is fundamentally proved by two observations. The first one is that the least squares means (LSM) obtained by varying the number of polygons and keeping constant the number of vertices are substantially similar (16.6, 17, 16.2). The second one is that the ANOVA test calculates an F-Ratio (1.1) is smaller than the corresponding F-critical ($F(2,120)=3.07$).

Graphically, Figures 3(a) (800x600) and 11 (1024x768) show that for each level of the *NV* variable (4) the *SSR* diagrams are essentially straight lines, which confirms the numerical results, namely for each level there is a substantial flattening of the means.

Thus, we can deduce that the *NP* factor does not affect the significant simplification rate. As a matter of fact, this assertion might be interpreted as the fact that the perception of significant changes is not affected by the inner complexity of the map or, we can claim that any set of maps having a similar number of points but with different number of polygons may be reduced using the same RDP rate obtaining no significant differences with respect to the original ones. From a practical point of view, as an example, it means that two maps of 100k vertices, the first one composed

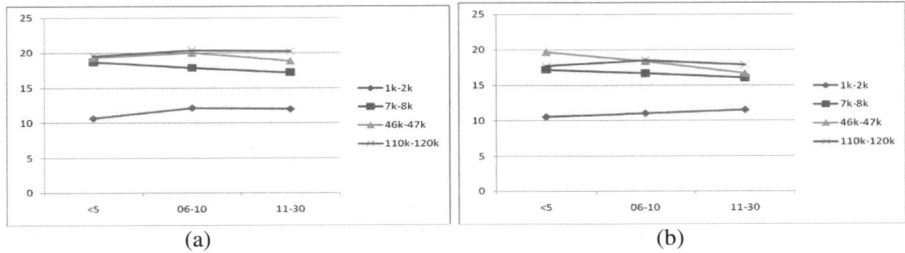

| (a) | (b) |

Fig. 3. (a)The diagram of the least square means considering NV constant and changing NP at 800x600 RES level, (b) The diagram of the least square means considering NV constant and changing NP at 1024x768 RES level

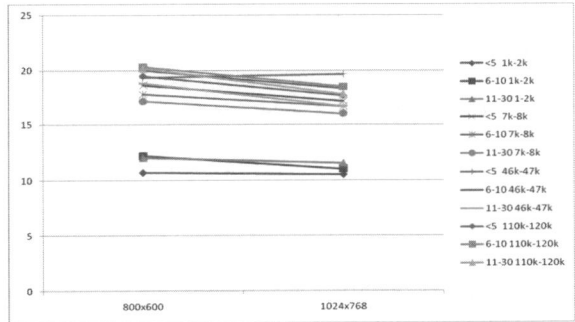

Fig. 4. The diagram of the least square means considering constant NV and NP and changing the RES levels

of one polygon while the second one composed by 100 polygons may be reduced using just the 30% of the original points without loosing significant details.

The last main factor we examine is *RES* (Hypotheses H3). Table 3 shows that the corresponding F-Ratio is greater than the critical F associated with the relative degree of freedom ($F(1, 120)=3.92$). Similarly to *NV*, it means we have to reject the null hypotheses, namely that in this case the variability is especially due to the effect of the *RES* factor. This claim may be considered highly significant because the p value is close to 0. As a matter of fact, The *RES* factor is relevant with respect to the simplification rate. Table 6 shows that *RES* decreases on the average from 17.3 to 15.9 passing from the 800x600 to 1024x768 resolution. It means that by increasing the screen resolution, subjects are more able to perceive significant differences between the original map and the reduced one. It is further proved by Figure 4. As a matter of fact, most of lines pass from a higher value to a lower one.

Table 6. Statistical about the *RES* factor

Factor	Level	LSM	SErr	N
RES	**800x600**	17.3	0.3	72
RES	**1280x768**	15.9	0.3	72

4 Conclusion and Future Work

In this paper, we have carried out an empirical study aimed at finding out which factors affect the user's perception of significant changes (SSR) after simplification using the Ramer-Douglas-Peucker algorithm. We have considered three factors, *NP*, *NV* and *RES*. We have proved that there does not exist significant interactions among any pair or triple of the above factors. Successively, we measured the contribution each factor provides to perception and we found out that just *NV* and *RES* give significant effects while *NP* does not. As for *RES*, the *SSR* decreases on the average from 17.3 to 15.9 passing from the 800x600 to 1024x768 resolution. It means that by increasing the screen resolution, subjects perceive earlier significant differences between the original map and the reduced one. Finally, as for *NP*, we proved that there are no significant differences by changing the levels in terms of *SSR*. As a matter of fact, it means that the complexity of the map in terms of number of contained objects is not relevant to the aim of map reduction.

In the future, we think to verify whether other kinds of factors affect the *SSR*. As an example, we want to perform experiments considering also the roughness degree, the size of the presented map and some other *RES* levels. Successively, we want to apply the results we found out to some of the application fields we showed at the begin of the paper. As an example, we want to build up a new layer which is interposed between users and map servers in order to speed up the transmission of maps automatically reducing the number of vertexes effectively transmitted over the network. We want also implement progressive map transmission based on the data we gathered and we do not employed in this experiment.

References

1. Bloch, M., Harrower, M.: MapShaper.org: A Map Generalization Web Service. In: Proc. of Autocarto 2006, Vancouver, Canada, June 26-28 (2006)
2. Buttenfield, B., McMaster, R.B.: Map Generalization: Making Rules for knowledge representation, Longman Group, UK (1991)
3. Cromley, R.G., Campbell, G.M.: Integrating quantitative and qualitative aspects of digital line simplification. The Cartographic Journal 29(1), 25–30 (1992)
4. Douglas, D., Peucker, T.: Algorithms for the reduction of the number of points required to represent a digitized line or its caricature. The Canadian Cartographer 10(2) (1973)
5. Dutton, G.: Scale, Sinuosity, and Point Selection in Digital Line Generalization. Cartography and Geographic Information Science 26(1), 33–53 (1999)
6. ESRI ArcView, http://www.esri.com (accessed 3.2. 03/06/08)
7. Hershberger, J., Snoeyink, J.: Speeding up the Douglas-Peucker line simplification algorithm. In: Proc. 5th Intl. Symp. on Spatial Data Handling, vol. 1, pp. 134–143 (1992)
8. Kazemi, S., Lim, S.: Deriving Multi-Scale GEODATA from TOPO-250K Road Network Data. Journal of Spatial Science 52(1) (2007)
9. Kazemi, S., Lim, S., Rizos, C.: A review of Map and Spatial Database Generalization for Developing a Generalization Framework. In: ISPRS Conference, Generalization and Data Mining (2004)
10. Lee, D.: Generalization within a geoprocessing framework. In: Proceedings of GEOPRO 2003 Workshop, Mexico City, November, pp. 1–10 (2003)

11. MacEachren, A.M.: How maps work: Representation, visualization, and design. Guilford Press, New York (1995)
12. McMaster, R.B., Shea, K.S.: Generalization in Digital Cartography. Association of American Geographer, Washington (1992)
13. Montello, D.R.: Cognitive map design research in the twentieth century: Theoretical and empirical approaches. Cartography and Geographic Information Science 29(3) (2002)
14. Montello, D.R., Golledge, R.: Scale and Detail in the Cognition of Geographic Information. University of California, California (1999),
    http://www.ncgia.ucsb.edu/Publications/Varenius_Reports/
    Scale_and_Detail_in_Cognition.pdf
15. Nakos, B.: Comparison of manual versus digital line generalization. In: Proceedings of Workshop on Generalization, Ottawa, Canada (August 1999)
16. Ramer, U.: An iterative procedure for the polygonal approximation of plane curves. Computer Graphics and Image Processing 1, 244–256 (1972)
17. Saalfeld, A.: Topologicaljy Consistent Line Simplification with the RDP Algorithm. Cartography and Geographic Information Science 26(1), 7–18 (1999)
18. Sheldon, M.R.: Introduction to probability and Statistics for Engineers and Scientists, 2nd edn. Academic Press, London (2003)
19. Slocum, T.A., Blok, C., Jiang, B., Koussoulakou, A., Montello, D.R., Fuhrmann, S., Hedley, N.R.: Cognitive and usability issues in geovisualization. Cartography and Geographic Information Science 28, 61–75 (2001)
20. SYSTAT ver.12., http://www.systat.com
21. Topfer, F., Pilliwizer, W.: The Principles of Map Selection. The Cartographic Journal 3, 10–16 (1966)
22. Visvalingam, M., Whyatt, J.D.: Line generalization by repeated elimination of points. The Cartographic Journal 30(1), 46–51 (1993)
23. Visvalingam, M., Williamson, P.J.: Simplification and generalization of large-scale data for roads - a comparison of two filtering algorithms. Cartography and Geographic Information Systems 22(4), 264–275 (1995)
24. Wang, Z., Muller, J.C.: Line generalization based on analysis of shape characteristics. Cartography and Geographic Information Systems 25(1), 3–15 (1998)
25. Wu, S.T., Márquez, M.R.G.: A non-self-intersection Douglas-Peucker Algorithm. In: Proceedings Brazilian Symposium on Computer Graphics and Image Processing, SIBGRAPHI XVI (2003)
26. Zhao, H., Li, X., Jiang, L.: A modified Douglas-Peucker simplification algorithm. In: Proc. of Geoscience and Remote Sensing Symposium (IGARSS 2001), Sydney, Australia (2001)

Location-Aware Reminders with Personal Life Content on Place-Enhanced Blogs

Hideki Kaji and Masatoshi Arikawa

Center for Spatial Information Science, the University of Tokyo
5-1-5 Kashiwanoha, Kashiwashi, Chiba 277-8568, Japan
{kaji,arikawa}@csis.u-tokyo.ac.jp

Abstract. One of the reasons users record their experience and knowledge as diaries, memos, photos, voice records and videos is to remember them for the future. However, the records are not effectively used in our real lives. The records are called personal life content in this paper. This paper proposes a location-aware personal life content reminder as a software tool to provide users with personal recorded content if they are located at the right places where the content was created or associated. The reminders are designed based on our developing place-enhanced Blogs which provide users with rich environment to tag personal life content with place descriptions. We discuss the architecture and the characteristics of both the location-aware reminders and the place-enhanced Blogs.

Keywords: personal LBS, life content, Blog, location based reminder, positioning descriptor.

1 Introduction

Commercial LBS (location based service) has become common in Japan. People often use GPS on the mobile phones, and access LBS to find their positions, to search POIs (points of interest), to generate itineraries of their trips using complex time tables of public transportation, and to navigate in the real world (Arikawa et al. 2007). The commercial LSB can provide users with additional functions of recording and accessing their personal spatial bookmarks and logs such as their favorite POIs, history of queries they made and results of the queries including computer-generated itineraries, and their trajectories. Then, users can use their spatial bookmarks and history to set the destination when they make a navigation query. Depending on locations of users, the order of spatial bookmarks can be adaptively changed to make it easy for users to select POIs. In the future, according to history of users, the commercial LBS could predict where the users are going and what they want to search based on the current locations and today's schedules of users. Thus, the current commercial LBS is advanced in the function of recording the histories and some spatial bookmarks, however the commercial LBS cannot record and retrieve user created content. A lot of Internet users are recording their experiences and knowledge on the network today. The users' created content is called *User Generated Content (UGC)*. According to Nardi et al. (2004), one of major motivations for blogging is documenting one's life.

M. Bertolotto, C. Ray, and X. Li (Eds.): W2GIS 2008, LNCS 5373, pp. 164–177, 2008.
© Springer-Verlag Berlin Heidelberg 2008

Moreover, "internet.com" and "goo research" have surveyed 1047 Internet users in their 10s to 60s on their attitudes toward Blog, 40.2% of them have owned one's Blog and 26.5% of them are keeping their Blog in Japan (Blog periodic research 2007). We have studied on if Blog can be integrated with LBS to make it open and personal on the Internet. This paper proposes a new framework of personal LBS using personal spatial data as part of users' daily life content which can be generated and shared on the Internet.

Location-based Reminder (LBR) is one of the types of location-based services using personal spatial content. LBR works as PIM (Personal Information Manager) by managing users' everyday tasks like meetings, contacts, documents, events, and so on, also it provides these tasks depending on location of users (Ludford et al., 2006). Dey and Abowd (2002) defined many types of context-aware reminders. It needed to set all of situations and information for delivering. comMotion (Marmasse and Schmandt, 2000) allows users to command the reminder with speech. PlaceMail (Ludford et al., 2006) has Web based interface for managing place information and spatial content, also it has special applications for mobile phones for controlling delivery timings and receiving messages. LBR can be useful to make use of the Blog records by pushing appropriate past Blog entries to users at right place and time when the users go nearby the places associated to the Blog entries. In common Blog, a user writes memos about some interesting places on the Blog, the user appreciates the records only on the Blog system by pulling entries on Web browsers, and then most of these records will be buried in dead storages. But if these records can be use as LBS content, users can use them more efficiently.

From these points, we propose Blog based personal LBS in this paper. Personal LBS can provide users with their personal information or services based on locations and their personal information which are managed in Blog and personal schedule management software. For instance, when a user is close to a bookstore, the personal LBS reminds the user to buy particular books which are recorded in the to-do list or the schedule with places to recall. We have realized the prototype system of the personal LBS based on our developing place-enhanced Blog which can integrate Blog, that is, a personal information management, and spatial information of places. Most of current Blogs realize the user environment to manage diary, schedule, to-do list, album and memos. We extend the Blogs to spatial information managing functions. We have also developed Web-based and email-based personal LBS on the mobile devices. Furthermore, users can write and read personal spatial information on the place enhanced Blog, and can be reminded appropriate Blog entries based on the location of the user through the mobile devices.

2 System Architecture and Applications

pTalk are a software family to realize our proposed framework of personal LBS. pTalk has the following five software components.

(1) **pLog** = place Blog: an extended Blog with functions of dealing with place descriptions.
(2) **pFeeder**/weblemail = place feeder: It allows users to create spatial content which should include place descriptions through the interfaces of web and email.

(3) **pGator** = place aggregator: It processes spatial queries, sets time and intervals of the queries, and transmits the results of the query processing to users through the pCatcher interfaces.

(4) **pCatcher** = place catcher: It is a front end software running on mobile devices to interact with users. **pCatcher** connects users and **pLog** by **pFeeder** and **pGator** on mobile devices.

(5) **pk** = place kicker or positioning descriptor or location agent: It positions a user or an object to make the location condition of the spatial query to **pGator**. Its examples are GPS, RF-ID, Wi-Fi based positioning system (Place Engine 2008), movement simulator and manual chooser of places on an itinerary.

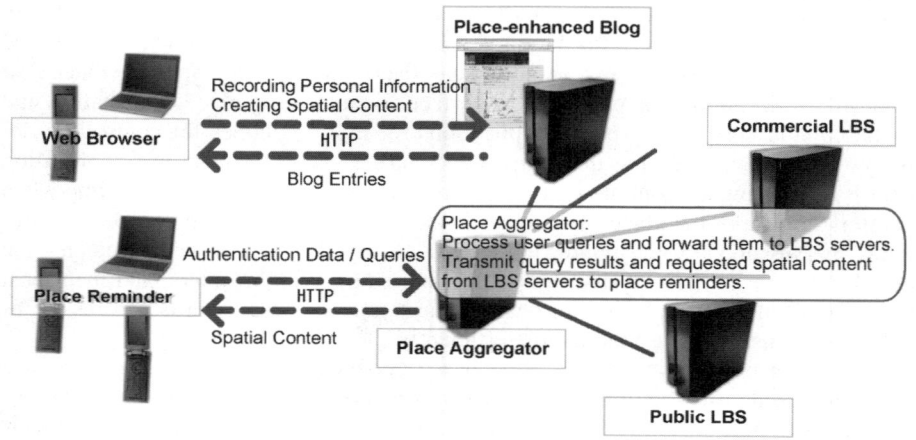

Fig. 1. Architecture of pTalk system

pTalk is a whole system composed of family software applications. Figure 1 shows a concept of the architecture of pTalk which is designed as an open platform to realize personal LBS based on protocols of Internet. Server side applications of pTalk include the place aggregator component and the place-enhanced Blog component. They are coded as Web CGI applications, and they communicate with client applications using HTTP. In addition, the place-enhanced Blog supports the email-based interaction with users through SMTP and POP3 protocols so that users can create entries of Blog and retrieve entries of their interest with the spatial conditions by creating commands on email messages. Place reminder is a client software of personal LBS on pTalk. It connects the place aggregator with HTTP to interactively retrieve and display POIs from the place-enhanced Blog through user-friendly GUI on the screen of a mobile device. Next two subsections show examples of use of our proposed system. The first one is a future diary for preparing of business trip and second one is a usage for collaborative education.

2.1 Future Diary: Setting Location-Based Reminders for the Future through pTalk

(1) Making an itinerary as future diary on place-enhanced Blog **pLog**

When we have some schedules, location-based reminders can be useful to automatically provide us with appropriate information at right places. Our proposed platform **pTalk** is suitable for such uses. Figure 2 shows an example of registering a schedule of a meeting at Hongo campus of the University of Tokyo on **pLog**. In advance, a user created the entry of the schedule on **pLog**. Examples of POIs created by the users are Hongo-Sanchome station, the places to have the meetings and passing points composing an itinerary from the station to the meeting room. The present **pLog** uses Google Maps to create and show POIs, itineraries represented by red lines on maps and diaries. (Figure 2)

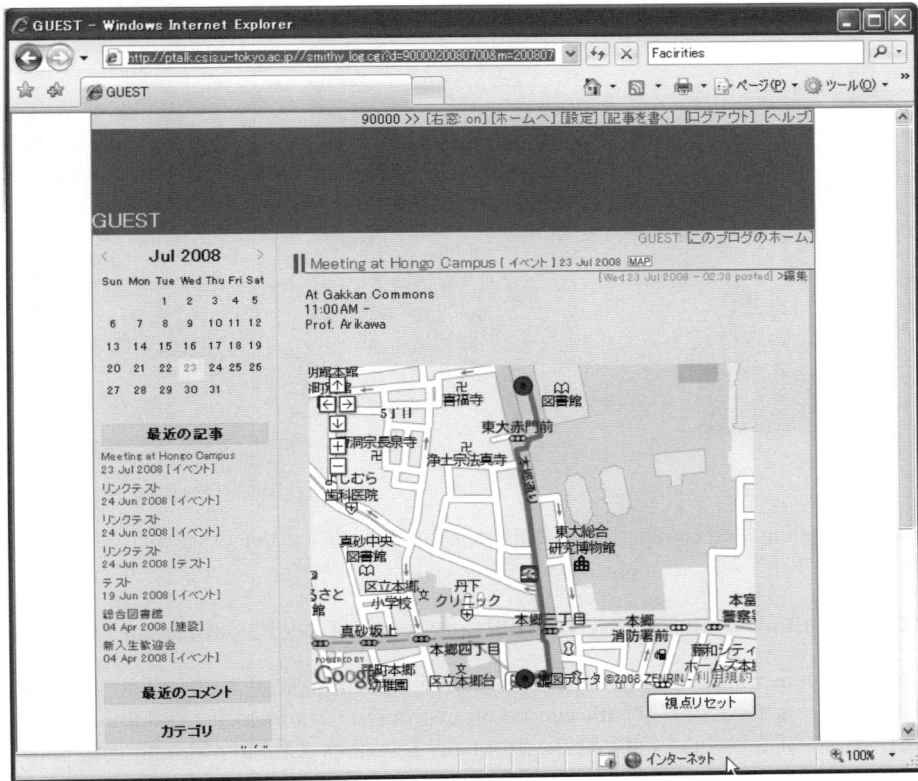

Fig. 2. A schedule entry of a meeting at Hongo campus of Univ. of Tokyo on **pLog**

(2) Personal LBS of the itinerary through **pCatcher** on a mobile device

When the scheduled date comes, we appreciate **pLog** entries pushed up to the location on pCatcher running on a mobile phone. An examples of a push service using a **pLog** entry on pCatcher is "When the user has arrived at Hongo-Sanchome Station, **pCatcher** shows him the itinerary of going to Hongo Campus of the University of Tokyo". **pCatcher** can also provide a map centered in his current location. The map can include additional POIs that are entries representing his past diaries. Furthermore, more POIs are available from entries created by other **pLog** residents and commercial LBS providers. Figure 3 shows an example of sharing POIs created by other **pLog** residents on the map on the mobile phone. The POIs is the content of Hongo campus guide created by a graduate student of the University of Tokyo.

Fig. 3. Recalling Blog entries, that is, POIs, depending on the location (Map data are provided by Zenrin Co., Ltd.)

2.2   Ubiquitous Collaborative Learning of Shapes of Big Fireworks for Children

Children often wonder if the shapes of big fireworks in firework festivals are the same from different locations. **pTalk** can be an appropriate platform for children to learn that shapes of fireworks look circles from any locations. Children should collaborate with one another, take photographs of fireworks from arbitrary positions by their GPS integrated mobile phone cameras, and send the photographs with email messages including the locations where they are taken and other related information to **pLog** through **pCatcher**. Children appreciate many photographs of fireworks taken from different locations by different children and locations of cameras taking the photographs on maps of **pLog** through Web browser (Figure 4). Red circle icons represent locations of cameras. The red circle icons may have their directions from cameras to the location of launching fireworks. The descriptions of directions of cameras can be indicated in the input form on mobile devices

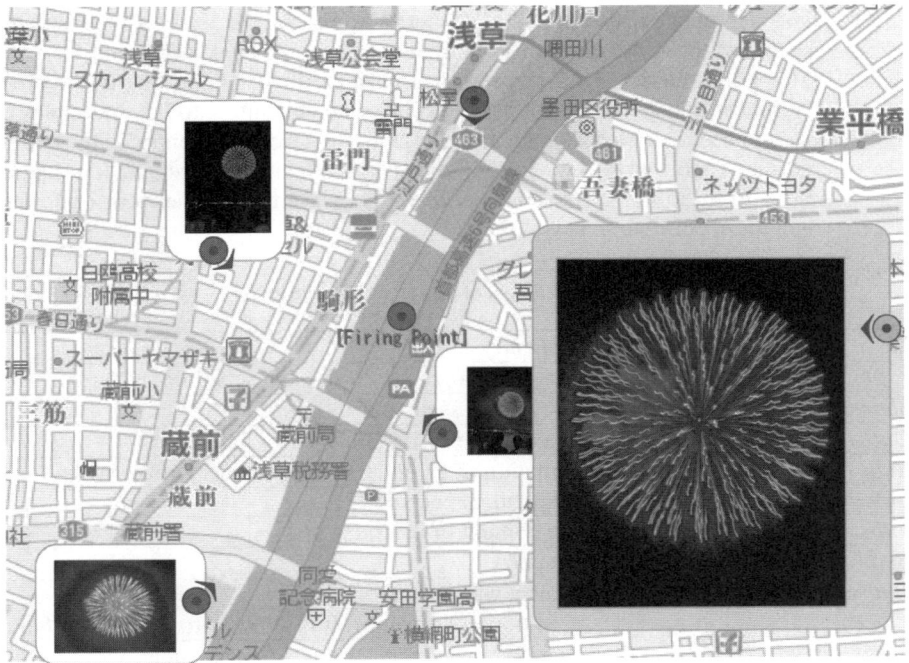

Fig. 4. Photographs of fireworks in a firework festival and their representative red circle icons attached with short and wide arrows are placed at the locations of cameras taking the photographs on Google Maps. These photographs are supposed to be uploaded by children using mobile phones integrated with GPS receivers and digital cameras.

3 Place-Enhanced Blog System *pLog*

A place-enhanced Blog system **pLog** provides users with an extended function of dealing with spatial information such as point of interest (POI) and area of interest (AOI) in addition to general functions of common Blog systems such as browsing and managing personal information. Users can add place descriptions to their Blog entries through the Blog input interfaces on Web browsers and email clients.

3.1 Entry of pLog

An entry of pLog is composed of the attributes listed in Table 1.

Some characteristic attributes of an entry stored in pLog are explained in detail as follows.

- *Type*: The *Type* attribute of all entries must be set to one of the selections including memo, diary, to-do and schedule. The *Type* attribute is used for calculating the importance levels for displaying Blog entries on a screen or a map. *Memo* entries are considered general information and less important in the order of visualization, because they do not have specific date attribute settings.

Diary entries have specific date attribute settings, but they are also considered less important in the order of visualization. They can be selected and set away by temporal queries. *To-do* entries are considered more important in the order of visualization. They do not have specific date attribute settings. Examples of *To-do* entries are places to go and things to buy. *Schedule* entries are considered much more important in the order of visualization if the specific date or time is close to the current date or time. They are treated the most important when the specific time is just now.

Table 1. Attributes of an entry stored in pLog

Id	Identifier for an entry
Type	Type of an entry (memo, diary, to-do, and schedule)
Shared group	Groups allowed to access an entry
Date	date when the events described in an entry happened
Author	Author's information of an entry (usually it is the Blog owner)
Category	category assigned to an entry
Title	title to brief the content of an entry
Body	main text of an entry
Transaction date	date when an entry was posted or edited
Place	A representative point of an entry. It is allowed to be null in blank.
Attached files	Media files attached to an entry, such as image, audio, video files

Table 2. Attributes of place object

Object ID	ID for a place object. It includes owner ID and 5 digit numbers.
Geometry type	type of geometry for a place object (0: Point, 1: Line, 2: Area, -1: Line point, -2 Area vertex)
Shared group	groups allowed to access an entry. They take over their parent entries' values.
Direction	direction of a place object. The semantics of the direction is flexible.
Latitude-Longitude, Zoom level	The position of a place object and the zoom level of a map displaying the place object are generated from the Web API of Google Maps.
Object name	name of a place object
Category	Category name assigned to a place object
[Alias name$_1$, number]	first alias name to the place object and total number of aliases
......	......
[Alias name$_n$, number]	n^{th} alias name to the place object and total number of aliases

- *Shared group*: The attribute *Shared group* is used for managing data access control. Special instances of the attribute values are "private" and "public" for entries to be prohibited and to be opened to the public. "friend" setting in the attribute *Shared group* is prepared for setting and creating groups for sharing entries. Users belonging to the group A can access entries of the group A.

- *Place*: All entries can be bound to the attribute *Place* settings. Some entries have Place settings, but others do not. Places are represented by single POI or higher dimensional geometries composed of multiple POIs.

3.2 Representation of Place Object in pLog

Our place-enhanced Blog pLog has functions to create graphic icons corresponding to pLog entries on the displace screen. All place information is represented as multiple POIs.

4 Creating Blog Entries with Place Descriptions by *pFeeder*

In this section, we introduce two software modules **pFeeder**/email and **pFeeder**/web to create pLog entries or Blog entries with place descriptions through email clients and Web browsers respectively. Both **pFeeder** can deal with assigning place information such as POI and AOI to Blog entries.

4.1 *pFeeder/Email*: Creating Entries through Email

pFeeder/email provides an environment for users to create Blog entries with place descriptions by email. The format of email messages for creating the entires is simple. We just write *UID* and *password* to the **Subject** field, and other attribute values such as *Title* and *Category* to the **Body** field in email messages. The email messages are sent to a specific email address, and they are automatically processed to create corresponding Blog entries in **pLog**. Attached files with email messages for creating Blog entries can be processed as *Attached files* of the Blog entries. An example of the content of the **Subject** field in the email message is as follows. "030903" and "doraemon" are values of the attributes *UID* and *password* respectively.

Subject: 030903, doraemon

An example of the content of the **Body** field in the email massage is as follows. They mean that values of the attributes *Title* and *Category* are "Rikyuu" and "sweet" respectively.

title="Rikyuu"
category="sweet"

Default value of the attribute *Shared group* is "private". Without explicitly setting values other than "private" to the attribute *Shared group*, the Blog entry never becomes public. If an image file attached with the email message includes place information in Exif format, the information is processed into geographic coordinate system and set to the attribute *Place* of the Blog entry. Thus, we can submit a Blog entry to **pLog** through **pFeeder**/email using GPS integrated mobile phone cameras.

4.2 *pFeeder/web*(1): Creating Place Tags Referring to POIs from Entries

It is almost the same to create entries on **pLog** as to create entries on common Blog systems through Web browsers. **pLog** provides functions to create *place tags* in

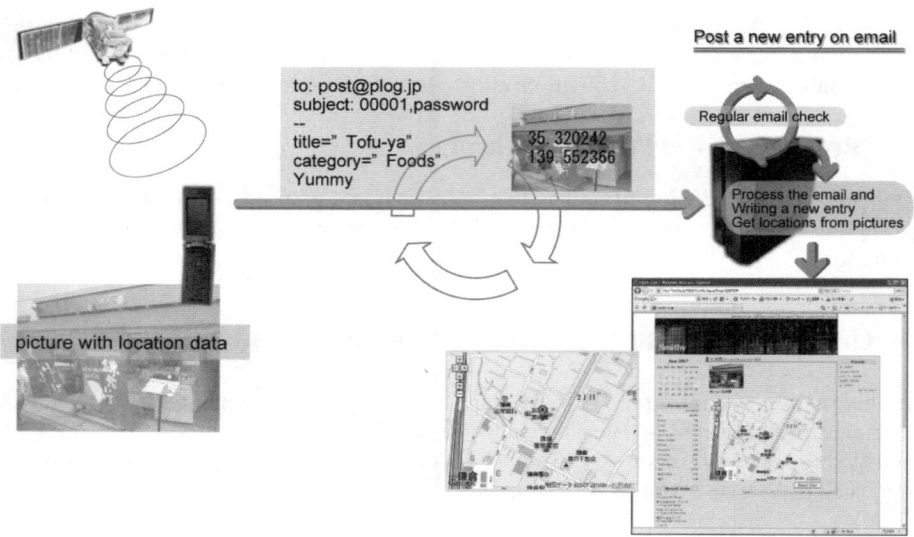

Fig. 5. Overview of **pFeeder**/email to create a Blog entry through email

entries. The following example shows that a spatial tag can be embedded in the body text of the entry. "poi231" and "Hanmoku-ya" means ID of POI stored in **pLog** and the name of the place respectively. The format of spatial tag in **pLog** is "[[poi*N*|*place_name*]]". The text including spatial tags can be translated into HTML text with hyperlinks and JavaScript by **pFeeder**/web.

> [Source text with a place tag]
>   "Today, I went to a new ramen noodle restaurant on Loop Line 2 at Hino.
>   Its name is *[[poi231|Honmoku-ya]]* and it serves Yokohama-style ramen noodle."
> [Translated executable text with a hyperlink]
>   "... Its name is <a href="javascript:jumpToPOI(231)">Honmoku-ya</a> and ..."

4.3 *pFeeder/web*(2): Creating Place Information on *pLog*

pLog provides users with three kinds of place icons on a map: point, line, and area. Line and area place icons are composed of point place icons. Point place icons can have additional direction decorations. **pLog** allows users to create and modify place objects in the database of **pLog** by creating and operating place icons on Google Maps (Figure 6).

Figure 6 is an interface of **pLog**. The followings are step-by-step explanation of using **pLog**.

(1) Choosing "Marker type".
    Location Marker has three types: Point, Line and Area.
(2) Moving the map and centering a target point.

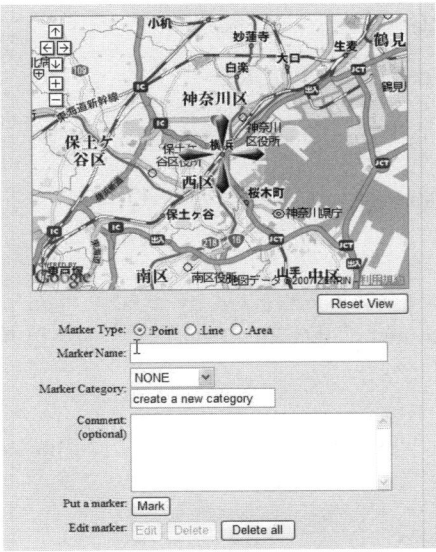

Fig. 6. Input form for creating place information through **pFeeder**/web

(3) Filling up input fields of place information, such as POI/AOI, Title, Category and Description. All the fields are not mandatory.

(4) Pressing the "Mark" button to generate a location marker on the map.

5 Recalling Entries Depending on the Locations by *pGator*

The place aggregator **pGator** aggregates entries and other spatial content related to the locations of users from multiple LBS servers, then it delivers the relevant content to users. **pGator** reduces the effort to search each LBS service. Users send queries to a pGator to search spatial content instead of searching multiple LBSs. Users make text-formatted queries to pGator to retrieve pLog entries chosen by the spatial conditions. The followings are an example of a query to a pGator.

    uid=09823
    pass=pass
    date=all
    place="139.7,35.7,1"

The above query means that Blog entries are created by the user "09823", and are located within 1km radius of the position at longitude 139.7 N, latitude 35.7 E. The followings are an example of the query result which is a XML text containing spatial content information, and satisfies the conditions of the above query.

```
<ENTRY>
 <UID>09823</UID>
```

```
<EID>0982320071102</EID>
<TITLE>Colloquium</TITLE>
<CATEGORY>Schedule</CATEGORY>
<DATE>8 Nov 2007</DATE>
<TEXT>At iii building 1F meeting room 10:00-</TEXT>
<POIS>
    <POINT>20071102,0,0,0,139.7610,35.7117,17,"Place","iii office",""</POINT>
</POIS>
</ENTRY>
```

The above example query is processed by pGator, then pGator returns relevant Blog entries from pLog, and also returns map image data of the relevant area generated from map servers.

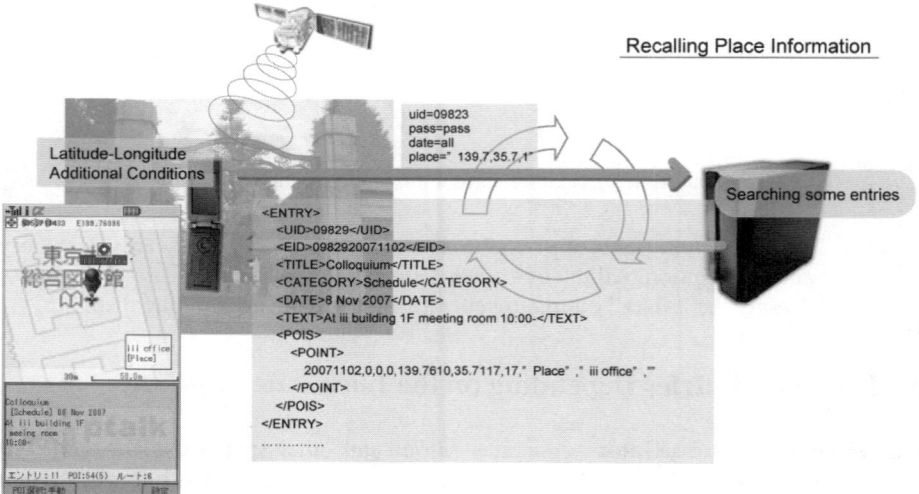

Fig. 7. Outline of **pGator** for pushing appropriate **pLog** entries to users depending on users' place requests on HTTP

Followings are examples of place requests for **pLog**. Example (1) describes that the user is located at the place "E139.4039,N35.3936", and wants to retrieve entries which satisfy the conditions of "date=2007.08.10-2007.08.12", "category=RESTAURANT" and "keyword=Italian, pizza". Example (2) shows the simplest setting of location and time for location-based reminder. The description of the example means that **pGator** is requested to push place information nearby the position "E139.6724,N35.8365" to the users at the time "2007.08.30,09:11:00". The place descriptor **pk** simulates the location of the user at the time by such description of place request. Examples (3), (4) and (5) describe that the location of the user is moving dynamically by the place descriptor **pk.** Example (3) uses a list of (position, time) for simulating a future trajectory of the user. Example (4) uses GPS to continuously position the user at default or user-determined setting intervals. Example (5) means that a location agent **pk** can use and choose any positioning devices adaptively to detect the most reliable position of the user. When we set a place request to **pGator** using email, a place request description like the following examples must be written in the **Body** field of the email message, and both UID and its password must be written in the Subject field of the email message.

> Example (1) of place request:
> **place=E139.4039,N35.3936**
> **date=2007.08.10-2007.08.12**
> **category=RESTAURANT**
> **keyword=Italian, pizza**

> Example (2) of place request:
> **place=*pk*(E139.6724,N35.8365,2007.08.30,09:11:00;)**
> **category=RESTAURANT**

> Example (3) of place request:
> **place=*pk*(E139.6724,N35.8365,2007.08.30,09:08:25;**
> **E139.6863,N35.7696,2007.08.30,09:08:55;**
> **E139.6865,N35.7002,2007.08.30,09:09:25;)**

> Example (4) of place request:
> **place=*pk*(GPS)**

> Example (5) of place request:
> **place=*pk*(any)**

6   Location-Based Reminder *pCatcher*

A location-based reminder **pCatcher** can provide a user with relevant entries of the user's **pLog** through a map interface (Figure 8). Circle icons correspond to entries of **pLogs** which have location attribute values near the current location of the user. The entries of **pLog** allow the user to remember her/his past memories which may become useful to go for the future. The example of a **pLog** entry is the content that a user has found an audio equipment shop which seems nice in the west area of Akihabara, but she/he has no time to visit there at that time. The user can browse the entry on Web browsers on mobile devises by choosing a circle icon and traversing its link. All digital content stored in **pLog** can be attached to place descriptions to be searched by spatial queries through **pGator**. The location of a user can be detected or simulated by **pk**. **pk** calls **pGator** to aggregate digital content within a certain distance radius of the location of the user. The distance of the radius can be changed depending on the scales for the use, the area size to display on the screen of a mobile device, the speed of the user, and so on.

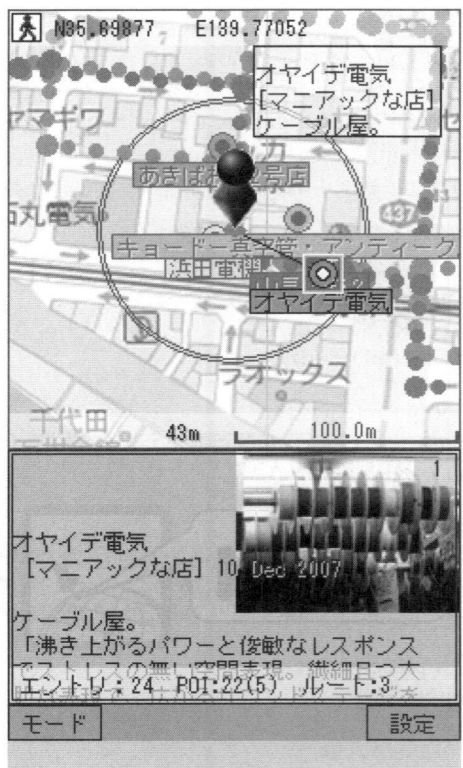

Fig. 8. An example of a screen of our developing location-based reminder software **pCatcher** running on a mobile device to recall spatial content stored in **pLog** by proximity to the current position of the user (Map data are provided by Zenrin Co., Ltd.)

7 Conclusion

Personal information managers (PIMs) on mobile devices can remind owners to do by alarm, voice, vibrator, flash light or other signals on specific time which the owners set in the schedule or the to-do list. Such digital reminders are designed based on time. On the other hand, digital reminders depending on location or place must be useful. For instance, when a user is close to a post office, the digital reminder makes her/him recall buying stamps. Most of the current PIMs are designed for dealing with only time, but not place to set things to do. There have been some researches and prototype systems for location-based reminders (Dey and Abowd 2002, Geominder 2007, Ludford et al. 2006, Marmasse and Schmandt 2000, Tarumi et al. 1999, Terada et al. 2001, Nair et al. 2006), but they are not common so far. There are many problems to solve for making the location-based reminders common, such as inaccuracy of positioning, tediousness of setting positions, complex user interfaces, dissemination of their usefulness, and marketing of them. Location-based reminders may be common in the near future on mobile phones or PDAs. Such location-based reminders are designed for organizing users' activities in the future. On the other hand, it may be good to auto-matically provide users with their private digital content such as diaries, memo, photos and past event in schedule recoded at the same place as they are currently located to remind their past activities. For instance, when a user walks around Akihabara down-town in Tokyo, some diaries, photos and past schedule are automatically displayed on his/her mobile device synchronized with places of the users and their content. Pushing users' past blog entries to each user can make them remember past forgotten memories and clarify their present situation from the life-span viewpoint.

We have proposed a personal LBS on place-enhanced blog in this paper. LBS are generally developed on the commercial telecommunication network services, and are usually not open in technical and use senses. Our proposal of personal LBS is open on the platform of Internet and the Web. Individuals can create and modify their own services for themselves by email clients, Web browsers and special software on mo-bile devices. The framework is easy to cooperate with commercial LBS. Then, users can use seamlessly both commercial and personal LBS.

References

Arikawa, M., Konomi, S., Ohnishi, K.: NAVITIME: Supporting pedestrian navigation in the real world. IEEE Pervasive Computing, Special Issue on Urban Computing 6(3), 21–29 (2007)

Blog periodic research, Daily research (in Japanese) (2007), http://japan.internet.com/research/

Dey, A., Abowd, G.: CybreMinder: A context -aware system for supporting reminders. In: Proc. Symp. Handheld and Ubiquitous Computing, pp. 172–186 (2002)

Geominder, http://www.ludimate.com/products/geominder/ (2007)

Ludford, P.J., Frankowski, D., Reily, K., Wilms, K., Terveen, L.: Because I carry my cell phone anyway: functional location-based reminder applications. In: Proceedings of the SIG-CHI conference on Human Factors in computing systems. pp. 889–898. ACM Press, New York (2006)

Marmasse, N., Schmandt, C.: Location-Aware Information Delivery with comMotion. Intl. In: Symposium on Handheld and Ubiquitous Computing, pp. 64–73 (2000)

Nardi, B.A., Schiano, D.J., Gumbrecht, M., Swartz, L.: 'Why we blog'. Communications of the ACM 42(12), 41–46 (2004)

Place Engine, Koozyt, http://www.placeengine.com/en (2008)

Tarumi, H., Morishita, K., Nakao, M., Kambayashi, Y.: SpaceTag: An Overlaid Virtual System and its Application. In: Proceedings of International Conference on Multimedia Computing and Systems (ICMCS 1999), vol. 1, pp. 207–212. IEEE, Los Alamitos (1999)

Terada, T., Tsukamoto, M., Nishio, S.: A Geographic Information System Using Active Data-base Systems. In: Proceedings of Symposium on Asia GIS 2001, CD-ROM (2001)

Nair, S., Kumar, A., Sampat, M., Lee, J.C., McCrickard, D.S.: Alumni campus tour: capturing the fourth dimension in location based notification systems, ACM Southeast Regional Con-ference archive. In: Proceedings of the 44th annual southeast regional conference, pp. 500–505 (2006)

Author Index

Printing: Mercedes-Druck, Berlin
Binding: Stein+Lehmann, Berlin